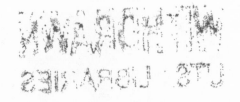

FLAVOR MICROBIOLOGY

FLAVOR MICROBIOLOGY

By

PINHAS Z. MARGALITH

Associate Professor of Applied Microbiology
Department of Food Engineering and Biotechnology
Technion-Israel Institute of Technology
Haifa, Israel

CHARLES C THOMAS · PUBLISHER

Springfield · Illinois · U.S.A.

Published and Distributed Throughout the World by

CHARLES C THOMAS · PUBLISHER

Bannerstone House

301-327 East Lawrence Avenue, Springfield, Illinois, U.S.A.

© *1981, by* CHARLES C THOMAS · PUBLISHER

ISBN 0-398-04083-4

Library of Congress Catalog Card Number: 80-12406

*With THOMAS BOOKS careful attention is given to all details of
manufacturing and design. It is the Publisher's desire to present books that are
satisfactory as to their physical qualities and artistic possibilities and
appropriate for their particular use. THOMAS BOOKS will be true to those
laws of quality that assure a good name and good will.*

Library of Congress Cataloging in Publication Data

Margalith, Pinhas.
  Flavor microbiology.

  Bibliography: p.
  Includes index.
  1. Flavor. 2. Industrial microbiology. I. Title.
TP372.5M37   664'.07   80-12406
ISBN 0-398-04083-4

Printed in the United States of America
C-1

to my teachers

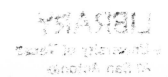

It is the same with wine when its bouquet has gone,
Or breath of perfume when its scent escapes in the air,
Or when from a substance flavor dies away,
And yet the thing itself does not seem any smaller to the eye,
Nor is anything lost in the weight;
Why? Because it is great numbers of small elements
Which make up smell, and give aroma to the
Whole substance of the things.

Lucretius, *De Rerum Natura*

PREFACE

I N PREPARING THIS TEXT, the author had in mind a twofold purpose. First, to emphasize the importance of microorganisms in the area of flavor formation in food technology, a subject frequently referred to but rarely organized under this title. Second, to expose the chemist, organic and biochemist, to a subject he may have had an opportunity to touch upon during his career but would prefer a larger coverage to which he or she may refer to.

This book is therefore intended for the microbiologist and chemist who are both deeply interested and often work in the field of natural products. Flavoring substances are natural products contributed not only by the composition of the food ingredients but also by the activity of various microorganisms taking part in the process of food preparation as its natural microbial population, or added intentionally to a specific food product during one of the stages of its processing. Obviously, food technologists will be the main workers to profit from this work. However, fermentation scientists, microbiologists, and others will undoubtedly gain satisfaction from the chemical diversity and biochemical complexity taking place in the work of microbes.

The biosynthetic activities of microorganisms have drawn the attention of microbiologists and chemists since the early days of microbiology. The complexity of these activities has become more and more evident with the introduction of modern analytical methods that reveal minute amounts of chemical substances concerned positively or that affect adversely the sensitive analytical tools of the human senses.

While devoting this work to microbiologists and chemists, it is assumed that the fundamentals of biochemistry are not unfamiliar to scientists of the two disciplines. Since a significant part of the subject treated refers to basic principles of microbiology to which many a chemist may not have had the opportunity to be exposed, a brief introductory chapter will be devoted to a general description of the biology of organisms contributing

vii

somehow to the formation of flavor in foods. A brief introduction to the physiology of the sensation of flavor will also be presented.

The contribution of microbes to the formation of flavor may be divided into several categories: (1) the enhancement of flavor, including taste and aroma, by the biochemical activities of a pure or mixed culture of certain group(s) of microorganisms; (2) the production of certain basic flavor components by organisms propagated under industrial fermentation processes and added to foods in a more or less purified form; (3) the undesirable effect of microorganisms on the formation of flavor. This is a somewhat extended field that comprises much of what may be designated as food spoilage microbiology. However, the last category will be dealt with only in terms of the formation of off-flavors, i.e. the production of minute amounts of chemical substance by microbes affecting profoundly the flavor of foods and their commercial potential without actually causing a gross spoilage of the specific product.

Many texts are now available that deal with the chemical activities of microorganisms. The now classic book by J. Foster, *Chemical Activities of Fungi*, which first appeared in 1949, was followed by the work by C. Rainbow and A. H. Rose, *Biochemistry of Industrial Microorganisms* (1963). Other titles dealing with the productivity of microbes from the pharmaceutical point of view (antibiotics, alkaloids and steroids) have appeared on the market. Most of these texts do not usually concentrate on the problem of flavor because much of the information on this subject has become available only during the last few years. While the early works deal with microbial products in terms of grams and percentages, flavor ingredients are also produced by microorganisms in ppm and ppb concentrations and have had to wait for the development of much finer analytical tools before their importance to flavor could be appreciated. An attempt has therefore been made to assemble much of this information under the present heading. Initially, it was decided to review research publications to 1977; however, while preparing the manuscript, several publications on the subject have appeared at a later date, and one could not resist their inclusion in the text.

Many chemical compounds described here are also given in structural forms. In this case, we usually followed the description available in the excellent treatise by S. Arctander, *Perfume and Flavor Chemicals* (1967).

Finally, I would like to remind the reader that the language used in this work is not the mother tongue of the author. I would like, therefore, to take the opportunity to apologize to the reader if a phrase used in this work does not express exactly what I intended to say. Also, I wish to express my gratitude to the Technion-Israel Institute of Technology, Haifa, which made it possible for me to devote some of my activities to preparing this manuscript.

Last but not least, this work is an attempt to describe, summarize, and present the work of many hundreds of research workers during the last few decades. Being scattered throughout many scientific journals and books, it is inevitable that some may not have come to my attention. However, in digesting this large amount of material, I have always been impressed by the hard work, thoroughness, and devotion of all who contributed to this field. It is they who made this book possible, and to them I wish to express my gratitude.

P.Z.M.

ACKNOWLEDGMENTS

IT IS A PLEASURE to acknowledge the assistance of many collaborators in preparing this manuscript: Mrs. L. Feld, our librarian, without whom the literature retrieval would not have been possible; Mrs. A. Katzenelson and Mrs. H. Nesher, who did the typing; Mr. A. Friede, who took care of the diagrams; and many other friends who lent me a hand. My sincerest thanks to them all.

CONTENTS

FLAVOR MICROBIOLOGY

. . . and it was like coriander seed,
white; and the taste of it was like wa-
fers made with honey.

EXODUS 17:30

Chapter 1

SOME MICROBIOLOGY FOR THE NONSPECIALIST

Microorganisms are among man's best friends and his worst enemies, but it took him a million years to find out.

—E. C. STAKEMAN

THE ORGANISMS

A
LMOST ALL ORGANISMS concerned with the production or transformation of food flavors belong to the plant kingdom of Thallophyta. These are characterized by being unicellular or multicellular organisms, nucleate or anucleate, with a photosynthesizing apparatus (chlorophyll) or colorless with a heterotrophic mode of nutrition, i.e. the consumption of ready organic materials. What they have in common is the absence of organized plant tissues so characteristic for the higher forms of plant life. Their cells or cell masses are therefore referred to as *thallus.*

Although many thallophytes may reach considerable sizes (the best known example would be the bulky seaweeds), most of the organisms related to food flavors have minute, microscopic dimensions. While bacteria (Schizomycetes) which appear in various shapes (rods, cocci, spirals) have a length or diameter of about 1 micrometer (micron, μm), fungi that are made of filamentous hyphae (branching filaments) are much larger. Hyphae have a diameter of about 10 micrometers and above. It is their small dimensions that impart such large surface area to these organisms, making them excellent catalyzers for many biological reactions in which their metabolism is concerned with the production of a multitude of compounds that also affect the organoleptic qualities of many foodstuffs.

Bacteria are procaryotic (anucleate) microorganisms that usually reproduce by binary fissions (Fig. 1) and under suitable

3

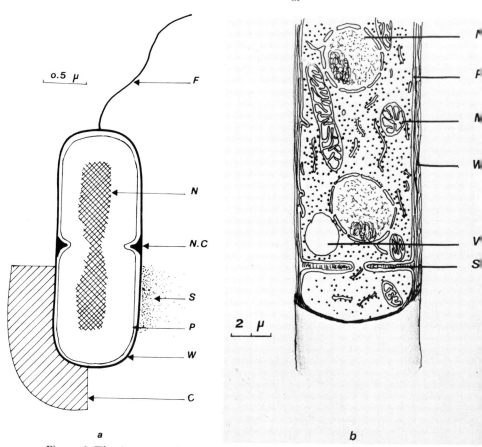

Figure 1. The Anatomy of Microbes (schematic). a. Bacterium. b. Fungus: C — capsular material, F — flagellum, N — nucleus or nuclear material, NC — new cell wall, M — mitochondrion, P — cytoplasmic membrane, S — slime, Se — septum, V — vacuole, W — cell wall.

environmental conditions (nutrition and temperature) may become visible to the naked eye in the form of a colony, when grown on a jellylike substrate, or as a turbid suspension when cultivated in a suitable liquid. Bacteria may have very short generation times (20 to 30 minutes) and reproduce rapidly in a logarithmic manner. A typical growth curve of a rapidly dividing bacterium is given in Figure 2. The logarithmic phase is usually preceded by a lag or adaptation phase and is followed by a

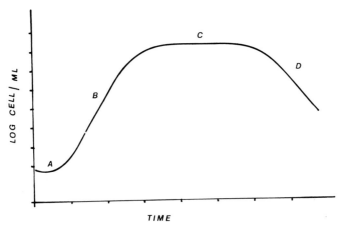

Figure 2. The Growth Curve of Bacteria. A — lag phase, B — logarithmic phase, C — stationary phase, D — decline.

stationary phase in which deacceleration of growth and sometimes also death take place.

When approaching the identification of bacteria, it is customary to determine their gram reaction first. This is done by staining the organism with gentian violet, decolorizing with alcohol, and counterstaining with a reddish stain. Some of the gram-positive bacteria, i.e. organisms that do not lose their violet color during the staining procedure, produce within each cell one heat resistant spore (endospore, Fig. 3). These spores are usually not killed by temperatures below 100°C, in contrast to the vegetative cells (bacteria without spores), which are readily inactivated by pasteurization temperatures (below 100°C). Without going into the anatomy and chemical composition of the bacteria, for which excellent textbooks are available, it is necessary to point out that in addition to size, shape, and gram reaction, the presence and number of appendages (mainly flagella), as well as their position in the bacterial cell, are also important features of bacterial taxonomy. The main orders of bacteria are usually recognized as follows:

1. Pseudomonadales — gram-negative bacteria; flagella, if present, are at the end of the cell; polar *(Acetobacter, Pseudomonas).*

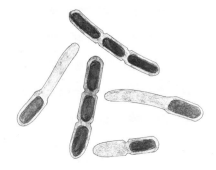

Figure 3. Bacterial Endospores (schematic).

2. Eubacteriales — gram-positive or gram-negative bacteria; flagella, if present, cover the whole cell, i.e. peritrichous *(Bacillus, Escherichia, Lactobacillus).*

3. Actinomycetales — bacterial organisms with hyphal-filamentous structures, reminding one very much of fungal mycelium but having a typical bacterial cytology *(Streptomyces,* Fig. 4).

Each of the bacterial orders may be further divided into families (e.g. Enterobacteriaceae), genera (the genus *Escherichia*) and species (the famous *E. coli*), and variants. In practice, not all organisms are suitably defined as to their taxonomic status. Expressions such as *Bacillus* sp., indicating that the organism has not been further defined below the genus taxon, are frequently found in literature. Microorganisms in general have a very large variability as to their metabolic and sometimes morphological nature. It is not uncommon that two identical cultures of equal species status show important quantitative or even qualitative differences. In this case, one usually designates the culture by a numerical label, e.g. *Lactobacillus brevis,* U 876.

Fungi are nucleate (eucaryotic) thallophytes and may be characterized by their mycelium consisting of branching filaments, the presence or absence of septa that divide the septate hyphae into individual cells, (see Fig. 1) and their mode of reproduction. While hyphae may grow in one plane (length), re-

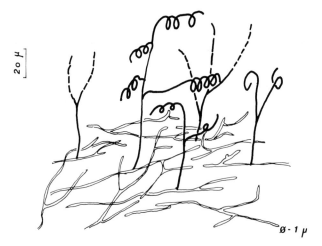

Figure 4. A Streptomyces (schematic).

production is usually carried out by special features — spores are produced by the breaking of certain parts of the mycelium into monocellular units or by the budding of special filaments into typical, often colorful, spore masses (conidia). The characteristic bluish color of the veins of blue cheeses are very well known examples. The formation of spores in special vesicles (sporangium) is also quite common (Fig. 5).

These spores are all the result of asexual reproduction. Sometimes fungi also display sexual processes in the production of their spores. As a result of the fusion between two haploid gametes, or their equivalents, a diploid cell is formed that is eventually followed by a meiotic process that yields again haploid cells. Sometimes such sexually derived spores may appear in readily recognizable structures that may harbor many millions of such spores. The best known example would be the delightful mushroom.

Contrary to bacterial (endo-) spores, fungal spores are not heat resistant and may be easily disposed of by low temperature heating, much like the vegetative hyphae. The classification of filamentous fungi is usually according to the following orders:

1. Phycomycetes —   aseptate fungi, with sporangiospores
                        (Rhizopus).

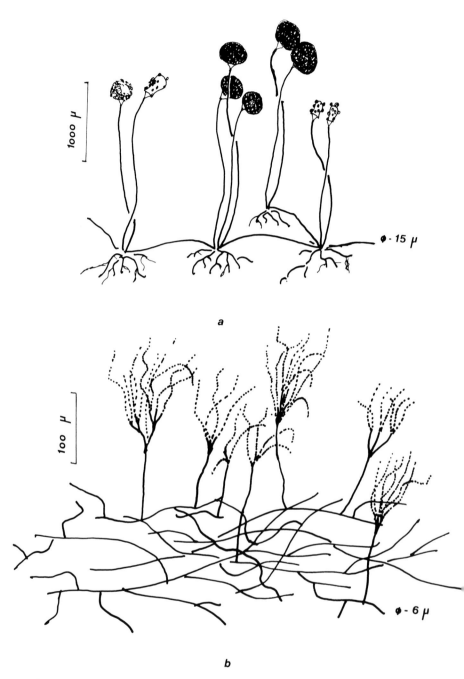

Figure 5. Mycelial Fungi (schematic). a. *Rhizopus* (sporangium), b. *Penicillium* (conidiophore with conidia).

2. Ascomycetes — septate fungi that also reproduce sexually with spores contained in a special structure: the ascus, hence ascospores *(Morchella, Tuber).*

3. Basidiomycetes — septate fungi that reproduce by sexual spores formed after fusion of nuclei in a special structure called basidium and the formation of spores upon these basidia, hence basidiospores *(Agaricus, Boletus).*

4. Deuteromycetes — or Fungi Imperfecti, a somewhat artificial group that contains ascomycetes or basidiomycetes, in which the perfect stage, the sexual phase, is yet unknown, making their final classification very difficult. Many of the filamentous fungi known to the public as moulds fall into this order *(Penicillium, Aspergillus).*

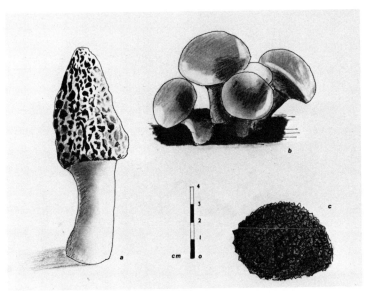

Figure 6. Mushrooms (schematic). a. *Morchella,* b. *Agaricus,* c. *Tuber.*

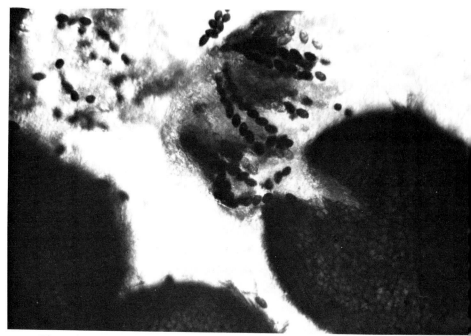

Figure 7. Asci and Ascospores. a. Fungus *(Sordaria)*, Perithecium with extruded Asci. Reduced 14 percent from ×400.

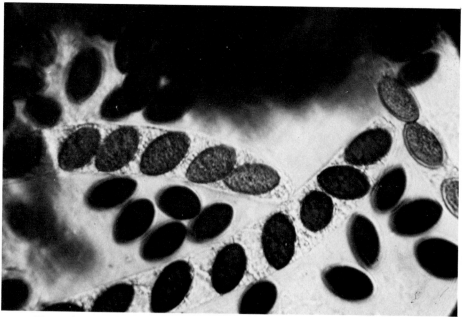

Figure 7b. Fungus *(Sordaria)*, Asci with Ascospores. Reduced 14 percent from ×1000.

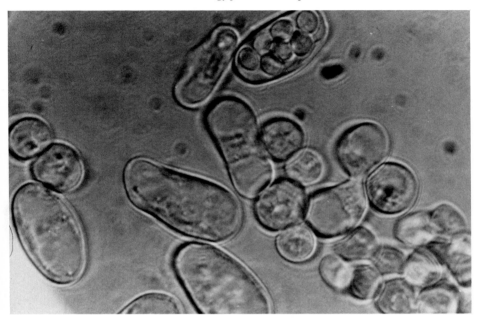

Figure 7c. Yeast *(Schizosaccharomyces)*, Ascus with Ascospores. Reduced 3 percent from ×1000.

Fungi may also assume nonfilamentous forms or may in addition to hyphae also produce unicellular structures that reproduce by budding or fission into other unicellular forms that may or may not become of filamentous character. In this case, we speak of yeasts or yeastlike organisms. The reproduction of yeast cells, e.g. bakers' yeast, only by budding, into unicellular forms would be the most typical case. There is good evidence that yeasts are a type of fungi since those yeasts that show sexual behavior are in general of the ascomycete type, i.e. they produce spores contained in the mother cell, the ascospores of yeasts (Fig. 7).

Algae are thallophytes that contain chlorophylls that enable them to fix CO_2 and build organic matter with the help of radiant energy. They may be unicellular or multicellular, anucleate (the blue-green algae or better cyanobacteria) or nucleate, with structures differing from unicellular motile *(Chlam-*

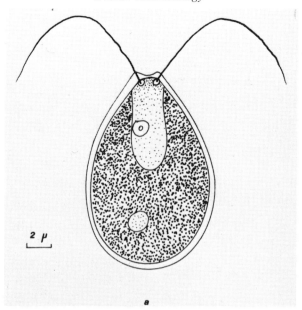

Figure 8. Algae (schematic). a. *Chlamydomonas.*

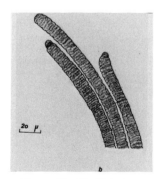

Figure 8b. *Oscillatoria.*

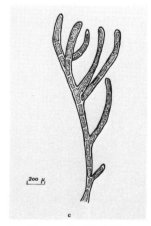

Figure 8c. *Cladophora.*

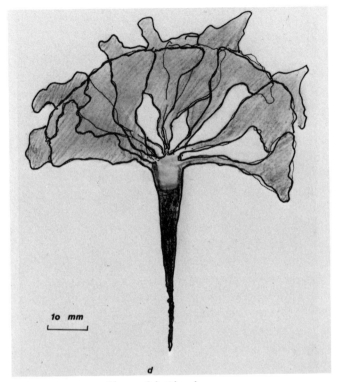

Figure 8d. *Chondrus.*

ydomonas) to filamentous forms like that of fungi *(Cladophora)* to the commonly known seaweeds. The taxonomy of algae is based largely on morphological considerations and on the chemical composition of the photosynthetic pigments, which vary considerably between the different groups (Fig. 8).

Algae are mainly sea and freshwater organisms but also occur on damp soil. We shall not go into more details with regard to their classification since we shall encounter them only occasionally in the consideration of certain flavor defects of fish.

SOME PHYSIOLOGY

Since nongreen thallophytes, fungi and bacteria, are the main groups of organisms we shall discuss with regard to flavor formation in foodstuffs, a short recapitulation of their major physiological characteristics seems desirable at this point.

Like other organisms, the heterotrophic thallophytes are made of organic material: carbohydrates, proteins, and nucleic acids built from suitable organic and inorganic sources — carbon (organic), nitrogen (organic or inorganic); phosphate, sulfur, and other minerals must also be included in the medium. Many microbes are unable to synthesize essential vitamins of the B group which must therefore be provided accordingly. Energy required for the metabolic process that enables the organism to grow and maintain its biosynthetic functions is provided by the oxidation of suitable carbon sources and the formation of high energy intermediates (ATP) that then drive the biochemical machinery.

The main feature of growth, after the nutritional and temperature requirements are fulfilled, is the supply of air. Many organisms among the bacteria are anaerobic, e.g. *Clostridium* sp., and develop best in the complete absence of oxygen; others are microaerophilic, that is they grow best at low oxygen concentrations. Many microorganisms are aerobic and require abundant air for growth. Of these filamentous fungi, most filamentous Actinomycetales as well as many Pseudomonadales are typical examples. Between the extreme aerobes and anerobes there are many organisms that can exist and thrive under both conditions, adapting their metabolic pathways to oxygen as the hydrogen acceptor during aerobiosis (respiration) or to other organic or inorganic compounds under anaerobiosis. The ability of many such organisms to switch from one type of growth to another is known as the Pasteur effect. The best known example is the adaptation of yeast cells to aerobic and anaerobic conditions. In the first, growth is accompanied by an increase of biomass (cell growth) with the production of CO_2 and water, i.e. the complete breakdown of the carbon source; while under *anaerobic* conditions, other products, e.g. ethanol, are formed along with abundant CO_2. The latter form is best known as *fermentation*. The production of lactic acid from carbohydrates by bacteria would be another example for fermentation. From thermodynamic considerations, it is obvious that the yield of growth under aerobic processes is far superior to the yields obtained under anaerobiosis.

Not all yeasts display the Pasteur effect; many are oxidative in their nature and behave just like other fungi in their relationship to oxygen, growing best on the surface of a substrate (film yeasts). Other yeasts have varying degrees of fermenting abilities, producing larger or smaller amounts of alcohol.

The term fermentation has also undergone important modifications during the last few decades. While the old concept of fermentation as described and studied by Pasteur is essentially an anaerobic process, now it is common practice to also include in this term aerobic processes, or in simple words, the growth of microorganisms on a large scale.

Since strongly aerobic organisms usually appear on the *surface* of a culture — which is, economically speaking, a poor process as much of the liquid is not utilized — a *submerged* culture process has been introduced which takes care of adequate air supply by shaking or forced aeration resulting in abundant microbial growth throughout the media (Fig. 9).

Much information is available today on the metabolic chain of reactions that leads to the formation of fermentation products by various organisms. The interested reader will find this information in current textbooks on biochemistry where many charts of metabolic pathways are also depicted.

In addition to nutrients and temperature, microbial growth is regulated to a large extent by the environmental hydrogen ion concentration. Most bacteria have an optimum pH between 6.0

Figure 9. Surface and Submerged Culture. Note fungal mat (right); tiny pellets in submerged culture (left).

and 8.0. While spore-forming bacteria will usually not develop at a pH below 4.5, non-spore-forming bacteria will also grow at lower pH values, even below 4.0, although these are extreme values which usually will not permit abundant growth. On the other hand, filamentous fungi, including yeasts and yeastlike organisms, have an optimum pH between 4.0 and 6.0 but will also proliferate at lower pH values. Since the pH of foodstuffs is of primary importance in their processing and preservation, the development of microbial growth and their biochemical activities, including the formation of flavor compounds, depend a great deal on the hydrogen ion concentration.

Another parameter that is of great importance in the maintenance of foods is their relative humidity (RH), or the more recent concept of water activity — A_w of foods. A_w equals P/P_0, where P = vapor pressure of a solution and P_0 = vapor pressure of pure water. A_w equals $RH/100$. Since water is quantitatively the most important ingredient of most living cells (vegetative microbial cells usually contain over 70% water), osmotic factors that regulate water availability from the environment will also regulate microbial growth. However, microorganisms differ a great deal in their susceptibility to limited water availability. While bacteria usually require very high A_w values (0.9 and above), fungi and yeasts have more moderate requirements, tolerating values down to 0.8 and below. But there are many important exceptions. Both bacteria and fungi may be osmotolerant and develop at lower values down to 0.65. However, there will be no growth at all below 0.6. The water activity of foods will be regulated by their water content and concentration of solutes. Generally salts and sugars are employed for this purpose.

Every microbial cell may be considered as a bag containing many thousands of different enzymes that control microbial life and activity. These mediate a variety of reactions that yield energy (respiration and fermentation) and perform the biosynthetic machinery of the cell. Although the energy yielding processes that are located within the cell are of primary importance in the growth of microorganisms and the products that are related to it (fermentation → alcohol), many reactions that are important in the production of flavor related compounds concerned with the organoleptic qualities of foods are also mediated through enzymes of microbial cells that are *excreted* into the environment. Of these, the most important are proteolytic, pectolytic, amylolytic, and lipolytic enzymes. Not all organisms have qualitatively or even quantitatively similar extracellular enzymes. Some are lipolytic while others are not. A good example would be the yeast employed for the fermentation of maltose derived from the starch in beer fermentation. Since such yeasts are practically devoid of extracellular amylolytic enzymes, other amylases have to be employed for this process. During food manufacture that employs prolonged curing processes (e.g. cheeses, certain meats), microorganisms play an important role in flavor formation, not only by extracellular enzymes but also through intracellular enzymes that are liberated upon cell death and autolysis.

HOW TO OBTAIN A SUITABLE ORGANISM

This is a subject of great interest to the applied microbiologist and will be dealt with here only briefly. Microorganisms may be isolated by either or both of two suitable elective methods: (1) a certain environment is provided that restricts the growth of unwanted organisms, (2) a certain environment is provided that encourages the proliferation of the desired type of organism. If one is interested, for example, in the isolation of a salt resistant, microaerophilic organism that can utilize malic acid as carbon source, one would set up a medium containing a suitable salt concentration together with malic acid in a closed vessel, inoculate with some grassy material or a grain of soil, and incubate at various temperatures. By repeating this procedure several times, an organism with such characteristics may eventually be isolated by streaking on suitable agar plates. It is much more difficult to isolate an organism for which no elective medium can be devised. In such a case, one needs a lot of patience and more often a good deal of luck to pick out the right organism from the millions that surround us.

Many processes, but not all, that are employed in the manufacture of foods and that depend on the formation of flavor by microbial activity do not involve pure cultures but rather the

Flavor Microbiology

combined action of a variety of organisms that develop as a result of a number of factors, such as ingredients, water activity, and environmental conditions. In this case, the study of each major group of organisms in relation to the rest may eventually lead to the improvement of the process. The general trend in microbiology of work with pure cultures has not always been successful in the food industry. Examples for such associative growth and action will be described.

FURTHER READINGS

Alexopoulos, C. J.: *Introductory Mycology*, 2d ed. New York, Wiley, 1966.

Metzler, D. E.: *Biochemistry, The Chemical Reactions of Living Cells*. New York, Acad Pr, 1977.

Pelczar, M. J., Reid, R. D., and Chan, E. C. S.: *Microbiology*, 4th ed. New York, McGraw, 1977.

Chapter 2

THE SENSATION OF FLAVOR

*Variety's the very spice of life
That gives it all its flavor.*

—W. Cooper

GENERAL CONSIDERATIONS

Two major features may be considered key parameters for the evaluation of foods: (1) nutritive value, (2) palatability. These characteristics of foods are interdependent, although in the everyday sense of food consumption, palatability and hence acceptance and appreciation will play the more dominant role.

Palatability can be viewed as being the sum of flavor, texture, and color. While both color and texture are mediated by microbial activity in some foods, especially during ripening, e.g. the textural changes that a Limburger cheese undergoes while the reddish pigmentation of its surface is underway, our main concern in this overview will be the effect of microorganisms on the flavor characteristics of foods.

The term flavor has been used in many different ways. In scientific literature, however, flavor is now generally considered to comprise the two major attributes, viz. taste (gustation) and odor (aroma, olfaction). Moncrieff (1951) defines flavor as "a complex sensation comprising taste, odor, roughness or smoothness, hotness or coldness, and pungency or blandness." Thus, to the two major categories of taste and aroma, a third one, viz. "feeling," may be added (Brozek, 1957). Whenever we ascribe certain taste or odor qualities to a chemical compound, it is good to remember that flavor is not the inherent property of a given material but rather the psychological attitude of the human senses (Hornstein and Teranishi, 1967). The integration in the human brain of the stimuli initiated by a chemical compound may be considered the ultimate sensation of flavor.

TASTE

Taste is a sense peculiar to the mouth and tongue. Historically, a large number of components of taste have been discerned, although today it is generally agreed that taste comprises only four basic taste characters, i.e. sweet, salty, sour, and bitter. However, modifications of these basic qualities are frequent; expressions of metallic, alkaline, and fatty tastes may be encountered.

The sensation of taste is achieved by the taste buds distributed all over the tongue and some restricted areas in the buccal cavity. Their number is estimated to be between 9000 and 10,000. Taste buds are located in tiny papillae in the anterior part of the tongue, whereas in the posterior portion they line the sides of minute trenches (Fig. 10). Each taste bud consists of about 10 to 15 taste cells from which minute villi protrude into the taste pore (Fig. 11). There seems to be a high turnover of taste cells, which are continuously formed by the cells surrounding the taste bud (Beidler, 1963). The taste buds are innervated by nerve fibers arising from the subepithelial plexus. About two-thirds of the tongue (the anterior part) is innervated by a portion of the seventh cranial nerve, becoming part of the chorda tympani nerve that enters the brain stem. The posterior part of the tongue is innervated by the glossopharyngeal (ninth cranial) nerve. For a detailed description of the anatomy of these tissues the reader is referred to a histological text, e.g. Farbman (1967). Although it was previously assumed that there exist different taste buds or receptor cells that respond to different taste stimuli according to the four basic taste qualities, this is apparently not the case. In fact, all taste buds respond to many of the taste modalities, although quantitatively they are not alike. There are special areas of the human tongue that are more sensitive to one of the basic taste characters. Thus, the sweet taste is best appreciated at the tip of the tongue, the bitter at the back, the sour at the edges, while the salty taste is best sensed both at the tip and at the edges (Moncrieff, 1967). However, it is the integrated taste response of a multitude of taste buds that creates the sensation of taste (Beidler, 1966).

There have been many efforts to explain the mechanism by

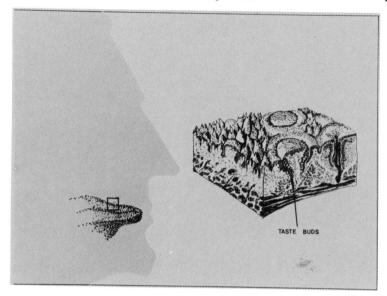

Figure 10. Portion of Human Tongue (schematic).

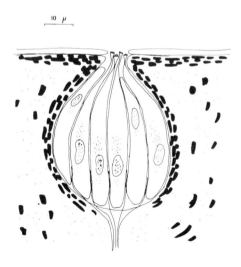

Figure 11. Drawing of Taste Bud. After Beidler (1952).

which taste active materials cause the sensation of taste. Beidler (1966) has shown that the initial step in the taste stimulation is a weak adsorption of the active material onto the receptor site on the surface of the taste cell. This in turn causes a depolarization in the charge of the nerve fiber leading to the formation of a nerve impulse. The involvement of enzymatic reactions in the creation of the taste stimulus has been suggested by Duncan (1963), although the absence of a temperature effect on taste sensation does not support such a possibility (Amerine et al., 1965).

One of the basic features in the evaluation of taste activity is the determination of the threshold value of the specific material, i.e. the minimum detectable concentration of the test substance by a suitable taste panel. This is a rather ill-defined term, since the *recognition* threshold is usually not identical with the *sensitivity* or *detection* threshold concentration. Thus, sucrose has a detection threshold of 0.17 to 0.55 percent versus a recognition threshold of 0.41 to 1.27 percent. However, in most cases, no separate data on threshold values are reported (Amerine et al., 1965). Threshold concentrations of different substances may differ by several orders of magnitude. Thus, an adult male will note sweetness caused by sucrose at a 0.5 percent concentration while the bitterness of quinine may be felt already at a concentration of 5×10^{-5} percent. Taste sensitivities depend greatly on the physical state of the taster but also on his age and sex. Thus, children have more sensitivity to the sweet taste of sugar (0.25% versus 0.5% of the adult), while women are more sensitive to sweetness and saltiness than males.

Since similar taste qualities may be achieved by compounds of different chemical structure, a brief discussion of the correlation between structure and sensation is pertinent. Sourness is achieved only by acids, most of their activity probably being due to the dissociation of hydrogen ions. Mineral acids cannot be distinguished from one another by taste, but their sourness is proportional to the hydrogen ion concentration. The relationship between sourness and pH is not so simple for organic acids. A solution of acetic acid tastes more sour than a solution of a mineral acid of the same pH, although the opposite is true when equimolar solutions of mineral and organic acids are compared.

Organic acids may also have tastes other than sour, e.g. citric acid also has a sweetish note.

The sensation of saltiness is produced by many low molecular salts, both cations and anions contributing to this sensation with different intensity. As the molecular weight of the salt increases, other qualities in addition to saltiness may rise, mostly bitterness. Thus, the saltiness decreases in the order: KCL, KBr, KI, while the bitterness increases accordingly (Moncrieff, 1951).

While sourness and saltiness are produced by distinct groups of chemical compounds, bitterness and sweetness occur in a large variety of unrelated compounds. From the chemical point of view, bitterness and sweetness seem to be very close. In fact, there are a number of compounds that are structurally very close but cause distinct taste qualities. For example, 2-amino-4-nitropropoxybenzene is about 4000 times sweeter than sucrose, while 2,4-dinitropropoxybenzene is bitter (Hornstein and Teranishi, 1967):

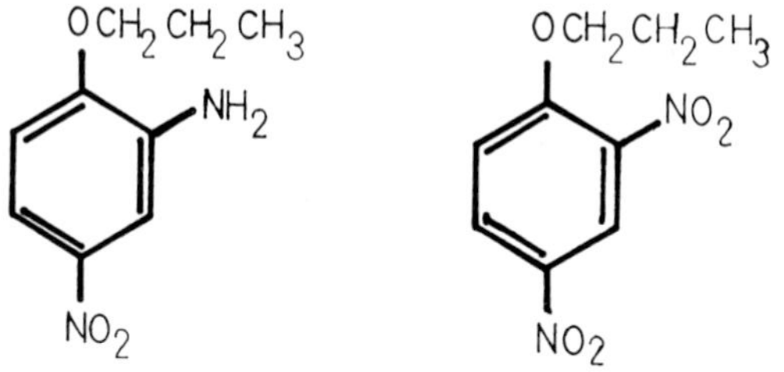

Another example is the commercially used sweetening agent, saccharin. A nitro derivate of saccharine is bitter:

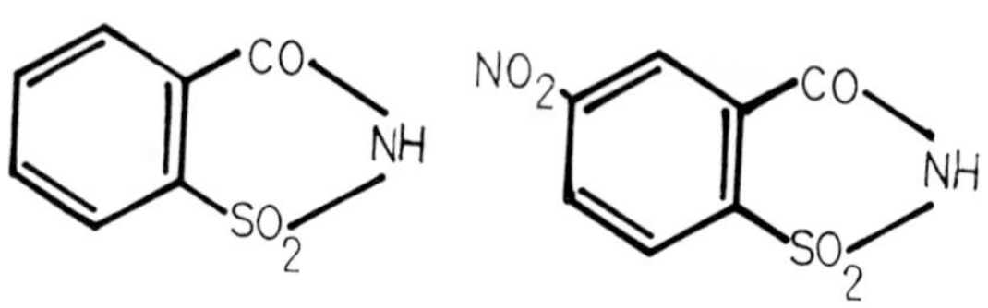

The closeness of bitterness and sweetness may be further deduced from the observation that an extract of *Gymnea sylvestra* leaves reduced both sweetness and bitterness but not saltiness, indicating a common step for the sensation of these two taste modalities. However, this observation was not confirmed by more recent work (Pfaffman et al., 1971). Isomeric changes also affect the taste sensation. Thus, o-nitrobenzoic acid is very sweet, m- is slightly sweet, while p- is bitter (Moncrieff, 1951). Optical isomers may have different taste qualities. The L-isomer of glucose is not sweet but slightly salty (Boyd and Matsubara, 1962). The anomers of mannose show a different taste. α-D-mannose is very sweet while β-D-mannose is very bitter (Steinhardt et al., 1962). Different tastes have been attributed to the stereoisomers of amino acids. For example, L-isoleucine is bitter, whereas the D-isomer is sweet (Berg, 1953). According to Shallenberger (1971), sweetness and bitterness are both the result of intramolecular hydrogen bonding as well as intermolecular hydrogen bonding of the taste molecule with the receptor site. According to this theory, sweetness requires two electronegative atoms such as oxygen about 3 angstroms apart, one with an available proton, the other as a potential proton acceptor for intermolecular hydrogen bonding with the receptor site. Bitter substances appear to have similar arrangements except that the proton to negative center distance is reduced about half, producing intramolecular hydrogen bonding and relative hydrophobicity.

The threshold value of flavoring material is considerably affected by the presence of various promoting or depressing substances. For example, sucrose has a pronounced effect on the threshold value of sodium chloride. In a mixture of both materials, the presence of sucrose at a concentration below 6 percent reduces the threshold value for salt, while above this value, the sensitivity to salt is diminished. On the other hand, the sweetness of sucrose was pronouncedly depressed by the presence of citric acid at very low concentrations (Pangborn, 1961). Clearly, the flavoring activity of a certain substance may be subjected to the enhancement or depression of other compounds even at very minute concentrations. In practice, a flavor enhancer may be defined as a seasoning agent that improves the flavoring properties of a particular food product, sharpening and emphasizing flavors *already present* without necessarily adding flavors of its own (Stier et al., 1967).

One of the early products of fermentation to be commercially promoted for the enhancement of flavor was monosodium glutamate (MSG). This product is now widely used in food technology to emphasize meaty and other flavors. The exact nature of the activity of MSG is still unknown, although several controversial hypotheses have been suggested (Amerine et al., 1965). The picture was further complicated when the flavor-enhancing activity of certain nucleotides was discovered. This may be due to a demasking action on certain taste receptors, thus exposing more sites to flavor sensation (Beidler, 1966). It seems that not only do these products affect foods by potentiating present flavor, but they also indirectly influence their taste activity by changing the viscosity of certain foodstuffs (Caul and Raymond, 1964).

Another interesting flavor potentiator is maltol, which has become commercially available in recent years (Harrison, 1970).

Maltol is now used in the confectionary trade to enhance the sweetness of sugars and synthetic sweeteners. It imparts more natural sweetness to synthetic sweeteners by masking an undesirable chemical tone when used at very low concentration (200 ppm and below).

Taste modification may also be of a more extreme character. A protein has been isolated from the miracle fruit *Synsepalum dulcificum*, which alters the sense of taste. If applied to the tongue, a sour substance becomes sweet. Thus, a lemon tastes like a sweet orange (Kurihara and Beidler, 1969). Obviously, the sensation of taste is an exciting area of human physiology and may be regarded as a promising field for new discoveries.

ODOR

The sensation of odor is of an even more intriguing nature. This is probably due to the fact that the perception of odor is much more powerful than taste with regard to threshold values. Thus, diethyl ether may be recognized at the low concentration of 1 mg per cubic meter of air, while vanillin is sensed even at a concentration of 10^{-7}mg/m^3 (Moncrieff, 1951). The different sensitivity of taste and odor may be demonstrated by ethyl alcohol, which has a threshold value of about 130 mg/liter water but may be smelled at a concentration of only 4 mg/liter air.

The higher sensitivity of odor may be due to the larger number of olfactory cells that are located in the yellow patch in the roof of the nasal cavity (Fig. 12). There are about 10 to 20 million cells from each minute cilia protruding into the mucus layer that bathes the nasal cavity. Axons from the olfactory cells constitute the olfactory nerve that leads to the olfactory bulb in the brain. Although humans perceive a large spectrum of odors, other mammals, e.g. dogs, have even more elaborate olfactory senses.

In the absence of any direct means for the measuring of odor, like taste but unlike light or sound which may be described in terms of wavelengths, the definition of odor is very difficult. Sagarine (1954) considers odor to be "the property of a substance that is perceived in the human and other vertebrates by inhalation in the nasal or oral cavity, that makes the impression upon the olfactory area of the body." In order that odor be perceived, molecules have to be accessible to the olfactory receptors, or in other words, a certain degree of volatility of the odoriferous material is implied.

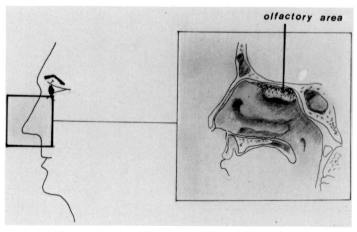

Figure 12. Cross Section of Nose (schematic).

Many theories have been put forward to explain the mechanism of olfaction. Today the most commonly held view is that of Davies and Taylor (1954) in which an odoriferous molecule (or odorivector) is adsorbed, dissolves in the lipid layer of the membrane wall of the olfactory neuron ending, and either by disordering its structure or by leaving a hole (membrane puncturing) that does not heal instantly, causes a collapse of the potential that exists across the wall of a neuron. This initiates the ionic flow and electrical impulse which propagates along the neuron.

The sensation of odor is further complicated by the fact that contrary to taste, an odorivector reaching the olfactory layer in the nose must (1) partition between air and the nasal mucosa (an aqueous phase) and (2) partition between the mucosa and the olfactory site membrane (probably a lipid phase). Amoore (1963) strongly emphasizes the importance of the stereochemical properties (size and shape) and electronic status of the material, believing in specialized receptors for each basic odor quality. In a more recent theory of odor perception, Randebrock (1971) proposes that the α-helix of peptides in the olfactory cilia is the major receptor site of odoriferous molecules. It is suggested that an oscillatory system could be formed within the helix. It is visualized that the odor molecule attaches itself by hydrogen bond formation with the amino acid residues of the helix, causing a modulation of the natural oscillatory frequency, transmitting to the nerve endings the modulated frequency. For a more detailed exposure of the theories of odor sensation, the reader is referred to the review by Dravnieks (1966).

The chemical nature of the odoriferous molecule is of great importance in odor sensation. Various saturated fatty hydrocarbons may have different odors depending on chain length; methane is odorless, hexane is easily distinguishable, and octane has a powerful gasoline odor. The amount of saturation also determines odor qualities since ethane is almost odorless, while ethylene is etheral (Beidler, 1966). In a homologous series, odors tend to reach maximum potency between C_8 and C_9, decreasing thereafter (Harrison, 1970). Information on odor differences of isomers (optical, geometric) is still not sufficient for making generalizations. Many observations have been explained by impurities of the test materials.

There have been many attempts to classify odors and to establish their components. Man can distinguish between an enormous array of different odors. Most authors agree that there are more primary odors than primary taste qualities. Amoore (1952) believes in only seven primary odors: (1) etheral, (2) camphoraceous, (3) musky, (4) floral, (5) peppermint, (6) pungent, and (7) repulsive. Other odors are considered to be the product of a combination of these primaries. Other authors have extended this list (Table I).

Mixtures of odors may depress certain major tones or create new blends (Sjostrom et al., 1957). The masking of undesirable odors by flavor-enhancing nucleotides is an unusual case of interaction between a taste promoter and a certain odor quality.

In dealing with flavor it is necessary to define the techniques of flavor evaluation. As previously stated, flavor is primarily the reaction of an individual to a certain compound via his sensory organs. Hence, the evaluation of flavor is greatly influenced by the person exposed to it. Age, sex, and culture, as well as mood and alertness, all affect the results of a certain flavor test.

TABLE I

BASIC ODOR DESCRIPTIONS ACCORDING TO VON SYDOW (1971)

Aromatic	Fragrant
Meaty (cooked)	Sweaty
Sickly	Almondlike
Musty, earthy, mouldy	Burnt, smoky
Sharp, pungent, acid	Herbal, green, cut grass, etc.
Camphorlike	Etherish, anaesthetic
Light	Sour, acid, vinegar, etc.
Heavy	Like blood, raw meat
Cool, cooling	Dry, powdery
Warm	Like ammonia
Metalic	Disinfectant, carbolic
Fruity (citrus)	Oily, fatty
Fruity (ether)	Like mothballs
Putrid, foul, decayed	Like petrol, solvent
Woody, resinous	Cooked vegetables
Musklike	Sweet
Soapy	Fishy
Garlic, onion	Spicy
Animal	Paintlike
Vanillalike	Rancid
Fecal (dunglike)	Minty, peppermint
Floral	Sulphurous

Several techniques are known for the sensory evaluation of flavor, some more suitable for a certain case than others. Generally, these tests are carried out by a panel of people employing one or more of the following methods: (1) difference, (2) rank order, (3) scoring, (4) descriptive, (5) acceptance and preference. For a complete discussion of these sensory tests in the evaluation of flavor, the reader is referred to a number of texts, e.g. Amerine et al. (1965), Moncrieff (1966), Goodall and Colquhoun (1967), Meiselman (1972). A statistical treatment of results is usually employed in order to reach valid conclusions.

The chemical analysis of odoriferous components of foodstuffs, which has made great advances since the introduction of highly advanced analytical tools, has prompted Guadagni and coworkers (1966) to introduce a new concept for the interpretation of flavor in foods. The term O_u, or odor unit, signifies the concentration of an odor component in a specific food divided by its threshold concentration. If the sum of the odor units of all the fractions in a mixture more or less equals the odor units of the mixture, then the contribution of each component or group of components may be expressed as *percentage of odor contribution.* If the sum of the fractions is substantially less or greater than the whole, the presence of suppression or a synergistic effect may be assumed. Such analysis of aroma profiles will be of great importance in the preparation of imitation products as was demonstrated by Salo et al. (1972) in whisky imitations. A similar approach to the contribution of taste components to the overall taste of food products should be attempted. In spite of the many basic facts that have accumulated in the study of the mode of action of odor perception, the physiology of olfaction is still poorly understood. The mechanism that enables the perception of substances at such low concentrations as 10^{-12}M or even 10^{-13}M still remains largely an intriguing mystery.

Coming back to the main topic of this text, it seems appropriate to recall the classic paper by Omelianski, who as early as 1923 collected information on the "aroma-producing microorganisms" — a designation commonly used today by many microbiologists. In his pioneering paper, Omelianski describes a large number of organisms related to the production of odor, from the sweetish scent that resembles lime tree flowers of

Pseudomonas pyocyanea to the pungent odor of sweat formed by a culture of *Bacillus fitzianus*. In an attempt to recover the aroma-producing properties of his culture, we read,

> The property of forming pleasant aromatic products represents a changeable character. To regain the lost characteristics, the culture is transferred upon the corresponding natural substratum: bacteria forming a strawberry aroma — on a decoction of strawberry leaves, bacteria with a pineapple aroma — upon a medium of pineapple leaves, bacteria with a radish-like odor — upon a medium of yellow radishes (Omelianski, 1923).

For the sake of piquancy, let us also mention an unusual case, when Omelianski isolated the so-called *Bacterium esteroaromaticum* from rabbit brain, which in a culture medium produced a pleasant fruity aroma resembling the odor of apples!

BIBLIOGRAPHY

Amerine, M. A., Pangborn, R. M. and Roessler, E. B.: *Principles of Sensory Evaluation of Food.* New York & London, Academic Pr, 1965.

Amoore, J. E.: The stereochemical specificities of human olfactory receptors. *Perfumery Essent Oil Record, 43:*321, 1952.

Amoore, J. E.: Stereochemical theory of olfaction. *Nature, 198:*271, 1963.

Beidler, L. M.: Our taste receptors. *Sci Monthly, 75:*343, 1952.

————: Dynamics of taste cells. In Zotterman, Y. (Ed.): *Olfaction and Taste.* Oxford, London, New York, Paris, Pergamon, 1963.

————: Chemical excitation of taste and odor receptors. In Gould, R. F. (Ed.): *Flavor Chemistry. Adv Chem Series, 56:*1, 1966.

Berg, C. P.: Physiology of the D-amino acids. *Physiol Rev, 33:*145, 1953.

Boyd, W. C. and Matsubara, S.: Different tastes of enantiomorphic hexoses. *Science, 137:*669, 1962.

Brozek, J.: Nutrition and behavior. In *Symposium on Nutrition and Behavior.* New York, Natl Vitamin Foundation Inc, 1957.

Caul, J. F. and Raymond, S. A.: Home-use test by consumers of the flavor effects of disodium inosinate in dried soup. *Food Technol, 18:*353, 1964.

Davies, J. T. and Taylor, F. H.: A model system for the olfactory membrane. *Nature, 174:*693, 1954.

Dravnieks, A.: Current status of odor theories. In Gould, R. F. (Ed.): *Flavor Chemistry. Adv Chem Series, 56:*29, 1966.

Duncan, C. J.: Excitatory mechanisms in chemo- and mechanoreceptors. *J Theor Biol, 5:*114, 1963.

Farbman, A. I.: Structure of chemoreceptors. In Schulz, H. V. (Ed.): *The Chemistry and Physiology of Flavors.* Westport, Connecticut, Avi, 1967.

Goodall, H. and Colquhoun, J. M.: *Sensory Testing of Flavor and Aroma.* Surrey, United Kingdom, Brit Food Manuf Ind Res Assoc, 1967.

Guadagni, D. G., Buttery, R. G., and Harris, J.: Odour intensities of hop oil components. *J Sci Food Agric, 17:*142, 1966.

Harrison, G.A.F.: The flavor of beer, a review. *J Inst Brew, 76:*486, 1970.

Hornstein, I. and Teranishi, R.: Chemistry of flavor. *Chem Eng News, 45:*92, 1967.

Kurihara, K. and Beidler, L. M.: Taste modifying protein from miracle fruit. *Science, 161:*1241, 1968.

Meiselman, H. L.: Human taste perception. *CR Food Technol, 3:*89, 1972.

Moncrieff, R. W.: *The Chemical Senses,* 2d ed., London, Leonard Hill, 1951.

——: *Odour Preferences.* London, Leonard Hill, 1966.

——: Introduction to symposium of foods. In Schultz, H. W. (Ed.): *The Chemistry and Physiology of Flavors.* Westport, Connecticut, Avi, 1967.

Omelianski, V. L.: Aroma-producing microorganisms. *J Bacteriol, 8:*393, 1923.

Pangborn, R. M.: Taste interrelationship. II. Suprathreshold solutions of sucrose and citric acid. *J Food Sci, 26:*648, 1961.

Pfaffman, C., Bartoshuk, L. M., and McBurney, D H.: Taste psychophysics. In Beidler, L. M. (Ed.): *Handbook of Sensory Physiology.* Berlin, Heidelberg, New York, Springer Verlag, 1971.

Randebrock, R. E.: Molecular theory of odor with the α-helix as potential perceptor. In Ohloff, G. and Thomas, A. F. (Eds.): *Gustation and Olfaction.* London, New York, Acad Pr, 1971.

Sagarine, E.: Odor: a proposal for some basic definitions. *ASTM Spec Tech Publ, 164:*3, 1954.

Salo, P., Nykanen, L., and Suomalainen, H.: Odor thresholds and relative intensities of volatile aroma components in an artificial beverage imitating whisky. *J Food Sci, 37:*394, 1972.

Shallenberger, R. S.: Molecular structure and taste. In Ohloff, G. and Thomas, A. F. (Eds.): *Gustation and Olfaction.* London, New York, Acad Pr, 1971.

Sjöström, L. B., Cairncross, S. E. and Caul, J. F.: Methodology of the flavor profile. *Food Technol, 11*(9):20, 1957.

Steinhardt, R. G., Calvin, A. D., and Dodd, E. A.: Taste-structure correlation with α-D-mannose and β-D-mannose. *Science, 135:*367, 1962.

Stier, E. F., Sayer, F. M., and Fergenson, P. E.: A comparison of methodology used in determining the flavor effect of 5-ribonucleotides on processed foods. *Food Technol, 21:*1627, 1967.

von Sydow, E.: Flavor: a chemical or psychophysical concept. *Food Technol, 25:*40, 1971.

Chapter 3

DAIRY PRODUCTS

*Hast thou not poured me out as milk and
curdled me like cheese?*

—Job 10:10

INTRODUCTION

DAIRY PRODUCTS are among the most popular foods and
constitute an important item in man's diet. No doubt milk
production and dairy industries occupy a prominent position in
agriculture in general and in food processing in particular. Dairy
products as we know them today, e.g. buttermilk, cream, and
cheeses, have evolved together with other food industries with a
high degree of know how and sophistication and with special
attention and care for the flavor aspects of these products. Basically, however, manufacturing of dairy products is derived from
principles that were already known during the early days of
civilization and may have their origin in nomadic life. The use of
animal intestines for pouches to carry milk and its subsequent
coagulation may be considered as the first step in man's discoveries in the field of milk preservation and technology. Many
of these observations have been applied through generations
even though the scientific principles underlying these processes
became clear only during the latter part of the nineteenth century. With the advent of modern chemistry, microbiology, and
nutrition, much information has accumulated that permits
modern technology to supply the public with dairy products of
high nutritive value, processed under strict sanitary regulations
and with flavor characteristics as preferred by the specific com-
munity.

From the microbiological point of view, the story of dairy
products can be described as the chapter of lactic acid bacteria.
These organisms have established an intimate association with
man's activity and health, which may have been motivated by his

32

need to preserve foods before conveniences such as refrigeration, heat treatments, or preservation by chemicals became available. In addition to dairy products, these organisms play an important role in the preservation of other foods by souring. The evolution of milk products with characteristic flavor profiles also turned out to be the result of the biochemical activities of lactic acid bacteria, although in many cases organisms from other taxa have also been involved.

The chapter of lactic acid bacteria is a very large one. We shall limit ourselves only to those that are related to the production of dairy products while many others involved in human disease will not be discussed, being outside the scope of this work.

The lactic acid bacteria or, in short, *lactics,* have the following general characteristics: gram-positive; non-sporeformers; microaerophilics, i.e. good growth when air excluded, not necessarily anaerobes; catalase-negative (no evolution of oxygen when brought into contact with dilute hydrogen peroxide; attack carbohydrates with production of varying amounts of organic acids, mainly lactic acid. The name lactics does not refer to the attack of lactose but rather to the production of lactic acid from carbohydrates.

The classification of lactics is based on their morphology — rods *(Lactobacillus),* or cocci appearing in chains with varying lengths *(Streptococcus),* in pairs or tetrades *(Pediococcus)* (Table II).

In addition, certain physiological characteristics are important in their classification. We distinguish between (a) homofermentative lactics, i.e. lactics that attack carbohydrates with the predominant formation of lactic acid (usually over 85%) with little or no other products, e.g. *Lb. bulgaricus,* and (b) heterofermentative lactics, which from sugars produce lactic acid and a variety of other products such as acetic acid, ethanol, and carbon dioxide in fairly large amounts, e.g. *Leuconostoc* sp. These differences in their metabolic products are related to different metabolic pathways. The homofermentatives employ the classical hexose diphosphate route of sugar degradation (the Emden-Meyerhof route), while the heterofermenters attack the carbohydrate via the hexose-monophosphate pathway, devoid of the key enzyme aldolase (Buyze et al., 1957). Although the lactic acid bacteria are widespread in nature, occurring both on vegetable material and substrates of animal origin, it is not easy to isolate them directly; usually an elective procedure is employed (suitable substrate with no air access) before isolation by plating is attempted. Media for the selective enumeration of cocci and lactobacilli have been suggested (Sharpe, 1955; Elliker et al., 1956). See Table III for an updated scheme for the differentiation of mesophilic lactic cocci.

Three of the aforementioned genera are predominant in dairy technology, i.e. *Streptococcus, Leuconostoc,* and *Lactobacillus.* The species of each genera have many important characteristics which will be described in more detail (Tables IV, V, VI).

In addition to sugar fermentations, temperature relationship, salt tolerance, etc., the optical activity of lactic acid (dextro-, laevo-, or meso-) formed is also considered an important factor in the classification of the lactic acid bacteria. We shall refer to this point in a later section.

Following the general layout of the morphology and physiology of these bacteria, it becomes evident that their major role in dairy technology may be summarized as (1) acid-forming bacteria: organisms that transform most of the fermentable carbohydrates (lactose) into lactic acid, and (2) citrate-fermenting lactics (the heterofermentative leuconostocs as well as the homofermentive *Str. diacetilactis*): organisms producing a variety of compounds, primarily diacetyl, that profoundly affect the flavor profile of a variety of dairy products. These bacteria are also referred to as *aroma bacteria* (Foster et al., 1957; Lindsay, 1967).

The lactic acid bacteria, although very active in their attack of carbohydrates, are further characterized by their poor biosynthetic capabilities. They are unable to grow in mineral media and require ample amounts of amino acids, purines, pyrimidines, and many of the B vitamins for optimal development. Although milk is usually considered as a rich medium for the growth of bacteria, the relatively low content of free amino acids (0.01%) makes it rather suboptimal for microorganisms such as the lactic acid bacteria that are unable to synthesize many amino acids (Reiter and Oram, 1962). Furthermore, lactic bacteria produce very little extracellular enzymes, have little proteolytic activity, do not attack polysaccharides, and possess com-

TABLE II
KEY TO IMPORTANT LACTIC ACID PRODUCING BACTERIA

Streptococcaceae		Leuconostoc (Heterofermentative)	Lactobacillaceae (Family)
Streptococcus	*Pediococcus* (Homofermentative)		*Lactobacillus* (genus) (Homo- and Heterofermentative)
cremoris	cerevisiae	citrovorum (cremoris)	acidophilus (species)
lactis		dextranicum	brevis
subsp. diacetylactis		mesenteroides	bulgaricus
thermophilus			casei
faecalis			lactis
			plantarum

TABLE III
SELECTIVE MEDIA FOR THE DIFFERENTIATION OF MESOPHILIC LACTIC COCCI

Organism	From	Media After	Selective Ingredient	Appearance on Selective Media
Leuconostoc sp.	*Str. lactis*	Galesloot et al. (1961) or Skean and Overcast (1962)	Ca-lactate + citrate	Leuconostocs and *Str. diacetilactis* appear in colonies surrounded by clear zones. Streptococci are inhibited by lactate.
Str. diacetilatis	*Str. cremoris*			
Leuconostoc sp.	Lactic streptococci	McDonough et al. (1963)	Tetracycline	Lactic streptococci are inhibited at 0.15 μg/ml TC or 75 ppm Na-azide
		Mayeux et al. (1962)	Na-azide	
Str. lactis	*Str. cremoris*	Turner et al. (1962)	Arginine – TTC*	*Str. lactis*, bright red colonies / *Str. cremoris*, white
		Reddy et al. (1961)	Arginine – BCP†	*Str. lactis*, white / *Str. cremoris*, yellow with yellow zone

* TTC = Triphenyltetrazolium.
† BCP = Brom cresol purple.

TABLE IV
KEY TO MAIN SPECIES OF THE GENUS STREPTOCOCCUS*

Species	Utilization of				Configuration of Lactic Acid	Growth at		Growth at NaCl		
	GLU	SU	LAC	Citrate		10°C	45°C	2%	4%	6.5%
lactis	+	−	+	−	L(+)	+	−	+	+	−
diacetylactis*	+	−	+	+	L(+)	+	−	+	+	−
cremoris	+	−	+	−	L(+)	+	−	+	−	−
thermophilus	+	+	+	−	L(+)	−	+	−	−	−
faecalis	+	+	+	+	L(+)	+	+	+	+	+

SOURCE: Adapted from Bergey (1974).

* According to the last edition of *Bergey's Manual of Determinative Bacteriology* (1974), this is considered only as a subspecies of *Str. lactis*. However, many bacteriologists prefer the spelling *S. diacetilactis* and consider it as a species because of the active citrate utilization and its inability (most strains) to produce ammonia from arginine (Collins, 1977). Nevertheless, *S. diacetilactis* maintains the serological properties characteristic of group N streptococci to which *S. lactis* also belongs (Swartling, 1951).

TABLE V
KEY TO MAIN SPECIES OF THE GENUS LEUCONOSTOC*

Species	Utilization of				Configuration of Lactic Acid	Growth at NaCl	
	Citrate	Lactose	Sucrose	Arabinose		3%	6.5%
cremoris (citrovorum)	+	+	−	−	D(−)	−	−
dextranicum	V†	+	+	−	D(−)	V	−
mesenteroides	V	+	+	+	D(−)	+	V
oenos	V	−	−	V	D(−)		V

SOURCE: Adapted from Bergey (1974).

* This genus has been earlier designated as *Betacoccus* (Orla-Jensen, 1919).

† V = variable with strains.

TABLE VI

KEY TO MAIN SPECIES OF THE GENUS LACTOBACILLUS

Species	Utilization of			Gas (Glucose)	Configuration of Lactic Acid	Growth at		Growth at NaCl		
	GLU	LAC	SU			15°C	48°C	2%	4%	6%
delbrueckii	+	–	+	–	D(–)	–	+		–	–
lactis	+	+	+	–	D(–)	–	+	+	–	–
leichmanii	+	+	+	–	D(–)	–	+		–	–
bulgaricus	+	+	–	–	D(–)	–	+	+	–	–
helveticus	+	+	–	–	DL	–	+	+	+	–
acidophilus	+	+	+	–	DL	–	+	+	+	–
casei	+	+	V	–	DL	+	–	+	+	+
plantarum	+	+	+	–	DL	+	–	+	+	+
brevis	+	–	–	+	DL	+	–	+	V	–
fermenti	+	–	V	+	DL	–	–	+	–	–
hilgardii	+	V	V	+	DL	–	–		–	–

SOURCE: Adapted from Bergey (1974).

paratively weak lipolytic enzymes. All these features make them extremely dependent on the type of media in which they develop (Reiter and Møller-Madsen, 1963). However, their comparatively feeble extracellular enzymic activity, prominent during short cultivation periods, is not negligible when extended incubations are carried out; under these circumstances, a large number of autolytic enzymes, liberated into the environment after the death of such cells, also become active. These enzymic activities assume a very important role in the process of curing (cheeses, meat products) when prolonged maintenance of foods at suitable temperatures and humidities is employed in order to obtain the desirable flavor characteristics of such foods.

Fermented dairy products, i.e. milk products that undergo any kind of microbial activity during their processing, occupy an important position in the dairy industry, both economically as well as from the consumer's viewpoint. These comprise various types of sour milk, yoghurt, sour cream, cream cheeses, cottage cheese, butter prepared from sour cream, as well as the numerous soft and hard cheeses. Although there are a large number of technologies concerned with the production of each of the dairy products, certain generalizations, especially with regard to their microbial aspects, may be made. (1) Short-lived products (liquid, semiliquid, or soft cheeses) are prepared by varying degrees of acidification, mainly due to the microbial activities of suitable starters prepared with the lactic acid bacteria. (2) Products intended for comparatively prolonged shelf lives, chiefly semihard or hard cheeses, are prepared from curds obtained through the enzymic activities of rennet or rennetlike enzymes (the so-called sweet cheeses). Lactic acid bacteria are frequently employed here in order to obtain slight acidification of the milk, which is advantageous to rennet activity, whey drainage, and flavor formation. However, in such cases the pH is much higher, never below 4.7, in comparison to the pH of around 4.0 obtained in the vats of sour cheeses. Also, rennet cheeses undergo microbial activities that contribute to their specific flavor profiles. Such activities are at least in part due to other types of microorganisms present or incorporated into the curd before curing. Such cheeses are also referred to as "Ripened Cheeses." (Day, 1967; Kosikowski, 1977.)

As we have seen, cultured dairy products (short-lived) differ from those of ripened products in many ways. From the point of view of flavor formation, there are no clear-cut boundaries since similar metabolic products may occur or are produced in both types. However, for the sake of clarity we shall also use this classification when dealing with the different processes that lead to the flavor development in dairy products.

THE FLAVOR OF CULTURED DAIRY PRODUCTS

Most of the lactic acid bacteria employed in dairy industries have been listed in Tables IV through VI. Although these organisms had already been described during the early years of bacteriology (Orla-Jensen, 1919), contrary to many other modern microbial processes, monocultures usually did not always replace the traditional use of mixed cultures. The use of monocultures often resulted in products inferior to those in which the traditional process was used. The reason for this phenomenon was not entirely clear for many decades, and only recent studies employing modern analytical techniques have explained the basis for the so-called "associative growth" of lactic acid bacteria. An example will illustrate this point. *Streptococcus cremoris* is the organism of choice for the production of lactic acid in dairies. This is usually accompanied by another closely related streptococcus, *Str. lactis,* which increases the efficiency of lactose utilization and the production of lactic acid (Table VII).

The production of folinic acid (citrovorum factor) by *Str. lactis* (Nurmiko, 1955) could account for this fact. Stimulation of growth and acid production by mixed lactic cultures may also be

TABLE VII
INTERACTION AMONG LACTIC STREPTOCOCCI

Organism	*ml 0.1 N NaOH/10 ml Culture*	
	Single	*Combined*
Str. lactis, F1A	2.24	
		7.15
Str. cremoris, F2A	4.83	

SOURCE: After Dahiya and Speck (1962).
NOTE: Cultures were prepared in reconstituted, sterile NFM (10% solids), inoculum 1.0 percent, incubation at 22°C for 18 hours.

due to other features. Dolezalek (1967) investigated the proteolytic activity of various pure cultures upon prolonged incubation in sterile skim milk and found that lactobacilli were in
general superior to streptococci in the production of dialysable
nitrogen compounds. Rapp (1969) described highest proteolytic
activity with *Lb. acidophilus* followed by *Lb. bulgaricus, Lb. helveticus,* and *Lb. casei.* The increased production of acidity by
streptococcal-lactobacilli mixed cultures is therefore not surprising.

Stimulation of growth and acid production may also be observed by various strains of the same species (Kothari and Nambudripad, 1973).

Further work with filtrates has clearly shown that the
stimulatory effect of *Str. lactis* was due to the excretion of significant amounts of adenine required by the fast growing *Str. cremoris.* Thus, milk cannot be considered as optimal medium for
lactic acid bacteria with highly efficient fermentation capabilities
(Dahiya and Speck, 1963). A synergism between thermophilic
lactics may be observed in Figure 13.

A similar example may be cited from Babel (1959) who examined the associative growth of *Str. lactis* with the aroma bacterium
Lc. citrovorum (cremoris) (Table VIII).

The associative growth of lactic acid bacteria is a frequent
phenomenon in dairy bacteriology. Many flavor nuances of
dairy products, as well as unwanted deviations from the desired
flavor profile, may be related to fluctuations in the associative
growth of these organisms.

Lactic acid is the principal flavor component of cultured milk
products. It is an odorless, nonvolatile acid that creates the
typical acid sensation of fermented dairy products. Since lactic
acid possesses an asymmetric carbon, two optical isomers as well
as inactive lactic acid are known to occur in lactic cultures. These
have little effect on the flavor profile of dairy products. However, nutritional studies have been conducted during recent
years, showing that the natural lactic acid for man is the $L(+)$
dextrorotatory acid. Human tissue and body fluids seem to
contain exclusively the dextrorotatory acid. Although the
human body can also use the levorotatory $D(-)$ acid, it
metabolizes the dextrorotatory lactic acid at a much higher rate

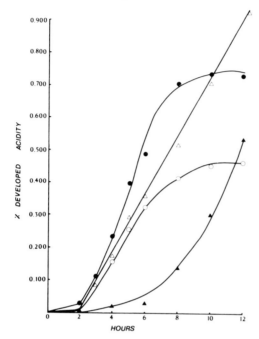

Figure 13. Commensalism and Competition in Mixed Cultures of *Lactobacillus bulgaricus* and *Streptococcus thermophilus*. Closed circle: mixed culture, open triangle: expected value (sum of developed acidity of both cultures), open circle: *Str. thermophilus,* closed triangle: *Lb. bulgaricus.* After Moon and Reinbold (1976).

TABLE VIII
SYNERGISM BETWEEN ACID-PRODUCING CULTURES

Organism	Hours				
	0	*2*	*4*	*6*	*8*
Str. lactis, alone	0.19*	0.19	0.23	0.36	0.64
Str. lactis + *Lc. citrovorum*	0.20	0.21	0.35	0.66	0.82
Lc. citrovorum, alone	0.19	0.19	0.19	0.19	0.19

Source: Babel (1959).
Note: Bacteria were inoculated at a level of 1 percent into skim milk and incubated at 86°F (30°C).
* Percent titrable acidity.

(Kupsch, 1972). These facts should be borne in mind when a choice for starter organisms is to be made since almost all streptococci produce the dextrorotatory L(+) lactic acid.

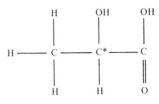

LACTIC ACID

While homofermentative lactics produce at least 85 percent lactic acid, heterofermentative organisms also produce considerable quantities of ethanol and/or acetic acid as well as CO_2.

The amount of lactic acid produced by the lactic acid bacteria depends on the substrate concentration (lactose concentration in milk may be increased by evaporation or addition of milk powder), incubation conditions and, of course, the organisms employed. Most of the homofermentative lactics produce abundant lactic acid (0.8 to 1.0%), even more than necessary for milk coagulation, which takes place at pH 4.6 to 4.7. In special cases, higher values of titrable acidity may be achieved employing cultures of *Lb. bulgaricus* or *Lb. acidophilus* (2% and above). Many of the high acid producing bacteria are thermophilic and require incubation temperatures of 43 to 46°C. Such thermophilic lactic acid bacteria have been used in processes for the production of yoghurt or when "cooking of the curd" is employed, as for many cheeses.

Lactic acid, although the major fermentation product of the lactic acid bacteria, is by no means the only metabolite. Even homofermentative lactics produce other products, some of which have a pronounced effect on the flavor profile of many dairy products. Among the volatile acids, acetic acid (0.010 to 0.015%) and to some extent propionic acid are produced by the homofermentative lactics such as *Str. lactis* and *Str. cremoris*. These acids do influence the aroma and taste of such starters. Interestingly, quantitative data of such volatile acids are scarce in literature, probably because attention to these acids has been

---

* Asymmetrical carbon.

directed to their production by heterofermenters. Recently it has been pointed out that the production of acetic acid by homofermentative lactics is strongly influenced by the state of aeration of the culture. While the microaerophilic condition is suitable for the predominant production of lactic acid, under vigorous aeration over 90 percent of the acids formed (0.2% glucose as carbon source) were acetic acid (Collins, 1977). Acetic acid is a terminal product of the hexose monophosphate pathway of the heterofermentative aroma bacteria (*Leuconostoc* sp.) and is also produced during citrate fermentation (Kandler, 1961). We shall consider these pathways in greater detail in the following section. Another heterofermentative organism should be mentioned here. *Bifidobacterium bifidus* (formerly *Lactobacillus bifidus*) has been isolated from stools of breast- and bottlefed infants as well as from adults. This organism differs from the classic lactic bacteria because of the irregular appearance and branching of their rods, and they are now included in the order of Actinomycetales (Bergey, 1974). Since this organism produces only the dextrorotatory L(+) lactic acid (up to 1.25% in milk cultures) together with considerable amounts of volatile, mainly acetic acid, its use as a starter organism for dairy products has also been suggested (Brown and Townsley, 1970). Other volatile acids such as formic, propionic, butyric, and valeric acids have been found in cultured milk products (Chou, 1962); however, because of their minute quantities, their role in flavor formation seems to be insignificant.

Among the nonacid C_4 compounds produced by the lactic acid bacteria, diacetyl (biacetyl; 2,3,-butanedione, dimethyl ketone or AC_2), acetyl methyl carbinol (acetoin, 3-hydroxy-2-butanone or AMC), and 2,3-butylene glycol (2,3-butanediol) occupy a special position in the flavor formation of fermented milk cultures. In spite of their minute quantities, their effect on aroma is very prominent.

DIACETYL   ACETOIN   2, 3-BUTYLENE GLYCOL

Diacetyl is a diketone with high aroma potential, especially in butter cultures but also in other fermented dairy products (including ripened cheeses) even when present in very small amounts. At low concentrations it is pleasing and definitely suggests the aroma of butter cultures, whereas at high concentrations it is pungent and rather objectionable (Hammer and Babel, 1943).

Acetoin, properly purified, has no odor, and at concentrations met in the dairy industry, probably has no effect on the taste of such products. Since its chemical structure is very close to that of diacetyl and since acetoin generally occurs together with diacetyl and at much higher concentrations, much thought and research has been devoted to its possible role as a precursor of the aroma principle diacetyl.

2,3-butylene glycol is the most reduced form of the C_4 group. The compound is also odorless when properly purified, and at concentrations found in dairy products does not affect their taste.

The heterofermentative aroma bacteria (*Leuconostoc* sp., especially *Lc. cremoris*) as well as the homofermentative *Str. diacetylactis* are outstanding in their ability to produce this group of aroma compounds. Our knowledge of the physiology and biochemistry of these organisms has greatly advanced during the last decade. New analytical techniques sometimes completely change earlier concepts about the biosynthesis of these organisms, thus providing greater knowledge about their metabolic activities and suggesting improved procedures to be applied in the dairy industry.

Since acetoin is usually formed together and in excess of diacetyl, for routine purposes these compounds were determined together. Under strong alkaline conditions (the Voges-Proskauer reaction) and in the presence of oxygen, a culture producing $AMC + AC_2$ produces a yellow color turning orange red within 2 hours. This may be accelerated to a few minutes by adding creatine and some α-naphtol to the test mixture (Ritter and Nussbaumer, 1963). Today, more reliable methods have become available which permit the quantitative measurements of each compound in milk cultures (Pack et al., 1964; Speckman and Collins, 1968a; Walsh and Cogan, 1974). It is because of the combined determination of AMC + AC₂ employing the Voges-Proskauer reaction that few analytical data of these compounds may be found in the earlier literature. In most cases, the combined values were considered to be a relative indication of the formation of aroma compounds.

Experience and bacteriological observations have taught the dairy microbiologist a number of facts concerning the production of aroma in milk cultures:

1. Aroma is seldom formed in appreciable quantities before acid conditions prevail in the culture. Even at 0.75 percent titrable acidity only one-third of the AMC + AC₂ was formed, as compared to 0.81 percent acidity (Barnicoat, 1937). Optimal pH conditions for aroma formation are from 4.4 to 3.8.
2. The presence of citrate is a prerequisite to aroma formation. Milk usually contains 0.15 to 0.18 percent citric acid. Addition of 0.15 to 0.20 percent citric acid to the media enhances aroma production by 100 to 300 percent.
3. Diacetyl formation is strongly related to oxygen availability (Branen and Keenan, 1971). No diacetyl will be formed when the culture is flushed with nitrogen or CO_2. The production of a highly flavored cream cheese type by growing aroma bacteria in skim milk under aeration has been the subject of a patent (Luksas, 1970a).
4. Optimal temperatures for aroma formation were 18 to 24°C.
5. Production of diacetyl also seems to be dependent on the temperature treatment of milk. This was demonstrated by Dutta et al. (1973) (Table IX).
6. Aroma compounds, especially diacetyl, have a transient nature and may decrease due to a variety of reasons (other bacteria, reducing conditions, etc.); hence, rapid cooling and aeration will preserve the diacetyl content (Pack et al., 1968).
7. Diacetyl production is the result of active metabolism of aroma bacteria. Chloroform killed bacteria will not produce AC₂ even in the presence of citrate or AMC.

The production of diacetyl under acid conditions may be

TABLE IX
EFFECT OF HEAT TREATMENT OF MILK ON
DIACETYL PRODUCTION (PPM)*

Temp.°C	Holding, Minutes	Str. diacetilactis (DRC₁)†	Str. thermophilus (Hst)
63	30	12	13
85	10	15	12
Steaming	10	9	6

* Adapted from Dutta et al. (1973).
† Cultures were examined in reconstituted nonfat milk after 18 hours incubation at 30°C and 37°C, respectively.

demonstrated with various acids, but in practice acidity is produced by the conventional lactic acid starters *(Str. lactis, Str.
cremoris)*. This is another important example of the synergistic
relationship between lactic acid bacteria. However, under these
conditions of rapid acid formation, the acid sensitive, diacetyl-
producing leuconostocs will propagate only poorly. Special procedures have been devised in order to ascertain high numbers of
leuconostocs in the culture before aroma formation sets in. In
general, aroma bacteria are less than 10 percent, in certain cases
only 1 to 2 percent, of the lactic microflora. The difficulties
encountered in the propagation of the leuconostoc aroma bacteria may be overcome by employing a mixed culture containing
Str. diacetylactis, a vigorous acid producer that synthesizes
diacetyl and other flavor components also at higher pH levels
with concomitant citrate utilization (Cogan, 1975). The use of
this organism has, however, some drawbacks — erratic flavor
production, excessive CO_2 production leading to a "floating
curd" undesirable in cheese production, etc. (Collins, 1977).
Attempts to improve the performance of *Str. diacetylactis* by
induced mutations have been recently described (Kuila and
Ranganathan, 1978).

The more recent trend of using concentrated, frozen, or
lyophilized starter cultures for various dairy products has raised
the question as to the effect of such cultures on the flavor profile
of the respective products. Although not enough experimental
work has been published on this problem, it seems that in some

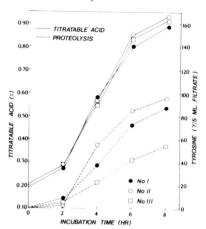

Figure 14. Titratable Acidity and Proteolysis Affected by Three Mixed Strain Starter Cultures Grown in Skim Milk at 32°C. After Williamson and Speck (1962).

cases even better results were obtained. Thus, Gilliland et al. (1971) claim that concentrated cultures of *Lc. citrovorum* have a greater capability for the production of diacetyl than conventional cultures.

The rate of acid production, as well as of diacetyl formation, may vary a great deal with different strains and mixtures occurring in commercial starter cultures. Slow acid producers — 48 hours of milk coagulation at 21°C versus 18 hours of fast strains (Citti et al., 1965) — usually have a more delicate, pleasing flavor often preferred by consumers of cultured buttermilk. Cultures that produce acid rapidly tend to have a flavor that is somewhat coarse compared to the slow cultures but may be preferred for the production of various cheeses. Some authors believe that fastness is due to increased proteolytic activity that permits more rapid growth of these exacting cultures (Williamson and Speck, 1962) (Fig. 14).

BIOSYNTHESIS OF AROMA COMPOUNDS

Flavor organisms bring about a double fermentation: (1) the standard sugar fermentation yielding lactic acid, (2) the citric acid fermentation giving rise to carbon dioxide, acetic acid, and the C_4 compounds. However, aroma bacteria do not utilize citric acid primarily. They grow very poorly on citrate alone but give a strong evolution of gas when both sugar and citrate are present.

The stoichiometry of C_4 production by the heterofermentative lactic acid bacteria has generally been described as:

$$\text{Citric acid} \rightarrow 1 \text{ to } 1.5 \text{ acetic acid } +$$
$$2 \text{ } CO_2 + 0.0 \text{ to } 0.5 \text{ } C_4 \text{ compounds.}$$

The higher the acidity, the more C_4 compounds are formed. This reciprocal relationship seemed to suggest a common precursor to the C_4 compounds and acetic acid. Pyruvic acid was considered to play the role of this intermediate compound (van Beynum and Pette, 1939). The structural similarity of the C_4 compounds and their difference in the oxidation state was considered for a long time to suggest their relationship with regard to their biosynthetic pathway. It was assumed that acetoin is the precursor of diacetyl, the former undergoing a biological oxidation. Pette (1949) was probably the first author to present evidence that diacetyl is *not* formed by biological oxidation of acetoin. He suggested the existence of a common intermediate, later identified as α-acetolactic acid, derived from pyruvate, which by oxidation in presence of atmospheric oxygen would produce diacetyl (DeMan and Pette, 1956). Since pyruvate occupies a key position in the formation of lactic acid via glycolysis, the occurrence of diacetyl only in the presence of citrate presented some difficulty. But diacetyl may also be formed in a sugar-containing mixture to which pyruvate is added. Hence, it is the excess of pyruvate that causes the shunt pathway leading to the nonacidic C_4 compounds (Mizuno and Jezeski, 1959).

The biosynthesis of diacetyl, however, turned out to be far more complex. In a series of brilliantly performed experiments employing modern biochemical separation methods and radioactive tracing, the Corvallis group investigated anew the biosynthesis of diacetyl (Seitz et al., 1963). The following steps were elucidated.

First, citric acid has to penetrate the bacterial cell. This was earlier shown to be facilitated by an inducible citric acid permease system (Harvey and Collins, 1962). After penetration, citritase, a constitutive enzyme not present in the homofermentative streptococci *Str. lactis* and *Str. cremoris* (Collins and Harvey, 1962), splits citric acid into oxaloacetic acid and acetic acid. The oxaloacetic acid is then decarboxylated to pyruvic acid ($+CO_2$). In the presence of thiamine pyrophosphate (TPP), it is further decarboxylated to acetaldehyde; 1 mole of activated acetaldehyde (acetaldehyde − TPP) condenses with another mole of acetaldehyde to form acetoin. Another molecule of acetaldehyde − TPP condensing with another molecule of pyruvate yields acetolactate which then gives another molecule of acetoin and in parallel also diacetyl.

Although the importance of acetaldehyde as a direct and indirect precursor of acetoin was clearly demonstrated, the biochemical basis for the oxidation of α-acetolactate to diacetyl was less clear. More recent work (Speckman and Collins, 1968b, 1973) has shown that another cofactor is essential for the formation of diacetyl without going through α-acetolactate. Acetyl-CoA seems to play a decisive role. It is probably condensed to OH-ethyl thiamine pyrophosphate (acetaldehyde − TPP) to yield the aroma principle. This was elegantly demonstrated when *Str. diacetilactis* or *Lc. citrovorum* were grown in a chemically defined medium devoid of lipoic acid (which is essential for the production of acetyl-CoA from pyruvate) so that acetyl-CoA could be formed only from added acetate which was suitably labelled. Under these conditions, diacetyl had a very high specific activity, while acetoin had little label. Hence, acetolactate was no longer involved in this reaction. The scarcity of acetyl-CoA in the system seems to be the bottleneck in the pathway that leads to the formation of diacetyl, most of the acetaldehyde giving rise to acetoin (Fig. 15). Some of the labelling found in acetoin is due to the irreversible reduction of diacetyl by diacetyl reductase. This reaction explains much of the erratic results obtained with aroma-producing bacteria, especially if such cultures are not suitably handled. This enzyme seems to be widely distributed in certain lactic acid bacteria but also in other groups. *Str. diacetilactis* is probably the most active of the lactics; others including *Lc. cremoris* are almost inactive, while many of the coliform bacteria (especially *Aerobacter aerogenes*) are very active (Seitz et al., 1963). Lower pH values favor diacetyl production by slowing down microbial growth, favoring citric acid permease

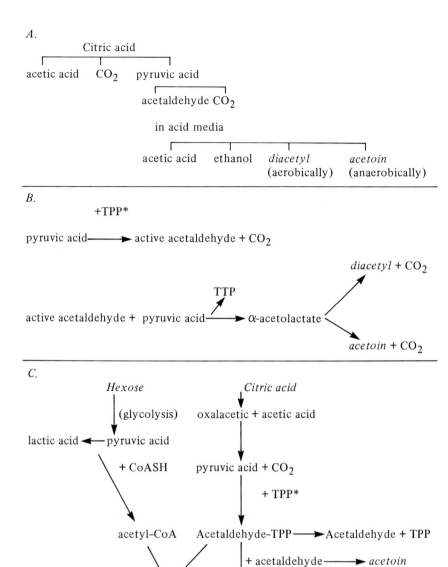

A.

Citric acid

acetic acid   CO$_2$ pyruvic acid

acetaldehyde CO$_2$

in acid media

acetic acid   ethanol   *diacetyl*   *acetoin*
                          (aerobically)   (anaerobically)

B.

+TPP*

pyruvic acid ⟶ active acetaldehyde + CO$_2$

                                                    diacetyl + CO$_2$

                                      TTP

active acetaldehyde + pyruvic acid ⟶ α-acetolactate

                                                    acetoin + CO$_2$

C.

Hexose   Citric acid

(glycolysis)   oxalacetic + acetic acid

lactic acid ◄— pyruvic acid

+ CoASH   pyruvic acid + CO$_2$

                  + TPP*

acetyl-CoA   Acetaldehyde-TPP ⟶ Acetaldehyde + TPP

                  + acetaldehyde ⟶ *acetoin*

                  + pyruvic acid

Diacetyl + CoASH+TPP

                  α-acetolactic acid + TPP

acetoin   *acetoin* + CO$_2$

*TPP – Thiamine pyrophosphate

Figure 15. Early and Recent Concepts of Citric Acid Fermentation by Aroma Bacteria. A. After Hammer and Babel (1943). B. After Seitz et al. (1963) and Stadhouders (1974). C. After Speckman and Collins (1973).

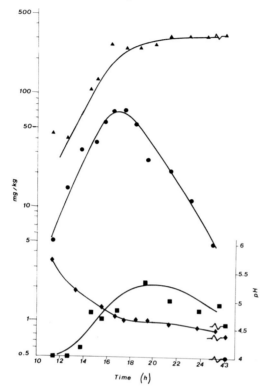

Figure 16. Fermentation Products in a Growing Starter Culture (D-type). Triangle: acetoin, circle: α-acetolactic acid, rectangle: diacetyl, diamond: pH. After Jönsson and Pettersson (1977).

(optimal pH below 5.5), and depressing diacetyl reductase (Collins, 1972). The further reduction of acetoin to 2,3,butylene glycol has only physiological significance without direct effect on the flavor system.

The general layout for the biosynthesis of diacetyl as contended by the California group has been confirmed by more recent work (Jönsson and Pettersson, 1977). Employing meticulous analytical procedures (maintaining an inert atmosphere during the distillation stage) it could be demonstrated that in starter cultures, α-acetolactic acid accumulates and is excreted into the surrounding medium. However, the environmental conditions that prevail during the propagation of a starter culture of aroma bacteria (low redox potentials) do not favor the spontaneous oxidative decarboxylation of α-acetolactic acid to diacetyl. Hence, under normal working conditions, α-acetolactic acid will not be the source of diacetyl, which is formed mainly as a result of the enzymatic reactions that take place within the cells. A representative analysis of flavor components in a starter culture is given in Figure 16.

This scheme, however, does not provide a comprehensive explanation for the importance of aeration for diacetyl synthesis. Bruhn and Collins (1970) believe that under aerobic conditions, oxidation of reduced nicotinamide adenine dinucleotide (NADH₂) to NAD⁺ by NADH₂ oxidase has a sparing effect on the reduction of pyruvate to lactic acid, thus leaving more pyruvate available for the formation of diacetyl. On the other hand, reductive conditions favor production of 2,3-butylene glycol from diacetyl by the AC₂ reductase, which prevents diacetyl accumulation. Hence, aeration will favor both formation and accumulation of diacetyl.

The complexity of diacetyl biosynthesis as illustrated by recent research constitutes another beautiful chapter of microbial biochemistry. The implication of many of the reported observations for the dairy bacteriologist are obvious.

QUANTITATIVE ASPECTS OF AROMA PRODUCTION

The more advanced analytical methods available for the determination of the C₄ compounds related to aroma in milk cultures now permit an evaluation of the quantitative results of these compounds in a variety of fermented dairy products. Here an additional point has to be discussed. Most of the work dealing with diacetyl formation was carried out with the characteristic aroma bacteria: the heterofermentative *Lc. cremoris* and the homofermentative *Str. diacetilactis* (although the latter produces diacetyl and abundant CO₂, it is definitely a homofermentative organism since CO₂ is produced only from citrate). First, is diacetyl production limited to these two aroma bacteria only? Many workers have presented evidence that other organisms comprising hetero- and homofermentatives are also capable of synthesizing diacetyl. Christensen and Pederson (1958) even claimed that homofermentatives produced diacetyl more readily than heterofermentatives (Table X).

Rushing and Senn (1960), working with concentrated orange juice (20° Brix) found strongest diacetyl production (after 20

TABLE X

PRODUCTION OF DIACETYL* IN CARBOHYDRATE SOLUTIONS

	ppm	
Organism	*Without Added Citrate*	*With Citrate (0.137%)*
Lb. plantarum	31.5	68.9
Lb. bulgaricus	66.0	75.5
Lb. brevis	0	0

SOURCE: Adapted from Christensen and Petersen (1958).
NOTE: Duration not given.
* Note high values due to analysis of total C$_4$ compounds, expressed as diacetyl.

days incubation, at 4 to 5°C) with *Lb. brevis* (up to 360 ppm), compared to 180 ppm of *Lb. plantarum*. They claim that *Lb. plantarum* var. *mobilis* produces diacetyl even in the absence of citrate or pyruvate. High values here were also due to the determination of C$_4$ compounds, given as diacetyl. Unfortunately, these results have not been confirmed by other workers. However, it has to be kept in mind that these investigations were performed in vegetable juices that developed a buttermilk off-flavor and may therefore be more related to vegetable pickling (see Chapter 5) than to milk cultures in which the formation is being followed during much shorter time intervals. Diacetyl was examined in milk cultures inoculated with *Lc. cremoris* (Gilliland et al., 1970) which after 6 hours reached levels of 2.7 ppm (10^8 cells per ml) in the presence of citrate. Highest levels (8.0 to 8.4 ppm) were obtained within 2 hours when the inoculum was derived from concentrated frozen cells. Branen and Keenan (1971) studied the production of diacetyl by *Lb. lactis,* a

TABLE XI

DIACETYL PRODUCTION BY *LB. LACTIS* IN BUFFERED SOLUTIONS

	ppm	
Condition	*Still*	*Agitated*
Control	5	20
+ citrate (16 μmole/ml)	4	25
+ acetate (16 μmole/ml)	5	27
+ pyruvate (16 μmole/ml)	10	250

SOURCE: From Branen and Keenan (1971).
NOTE: 24 hours incubation at 30°C.

homofermentative bacterium present in considerable quantities in ripened cheddar cheeses, and emphasized the positive effect of aeration and citrate on diacetyl production in chemically defined mixtures (Table XI).

Walsh and Cogan (1973) studied diacetyl production by lactic acid bacteria following the classification of Galesloot and Hassing (1961):

B-type starter containing *Leuconostoc* as aroma bacteria, up to 4 ppm diacetyl

D-type starter containing *Str. diacetilactis* as aroma bacteria, up to 12 ppm diacetyl

(cultures in 10% nonfat milk solids were incubated up to 14 hours at 21°C).

Also thermophilic streptococci incubated at 40°C were found to produce appreciable quantities of diacetyl. Of the cultures examined, 50 percent produced 2-3μg/ml diacetyl in autoclaved milk (Choisy, 1974). Drinan et al. (1976) have investigated production of C_4 compounds in a variety of lactic acid bacteria (Table XII).

There is but little published data on diacetyl content of commercial cultured dairy products and the effect of manufacturing procedures on these values. Mather and Babel (1959) examined diacetyl concentrations in forty-one samples of creamed cottage cheeses. Values ranged between 0. and 3.2 ppm, 95 percent of the samples containing less than 2.0 ppm. During cottage cheese manufacture, there is a separation of the milk constituents and flavor compounds, the whey fraction containing considerably more citric acids, lactic acids, acetoin, and diacetyl than the corresponding curd. However, a diacetyl content of about 2 ppm gives a pleasing flavor and seems to mask a sour flavor sometimes associated with cottage cheese. The best results are obtained by inoculating skim milk with a pure culture of *Lc. cremoris.* After sufficient growth has taken place, the pH is adjusted with sterile citric acid to 4.3. Further incubation at 21°C results in the formation of considerable diacetyl and volatile acid. The skim milk culture thus prepared is used to standardize the fat content of the cream to about 12 percent. Cheese cream with

TABLE XII
DIACETYL PRODUCTION IN MMRS MEDIUM

Organism	No citrate	ppm + 1.58 mg/ml citrate
Lb. viridescens 7-7	0	0
Lb. fermenti D-1	0	0
Lb. plantarum P5	0	2.1
Lb. plantarum G-1	0.7	1.9
Str. diacetilactis 1816	0.0	3.2
Str. diacetilactis DNCW4	0.0	6.1

Source: From Drinan et al. (1976).
Incubation at 30°C, diacetyl determination after 16 hours.

such a preparation has an appreciable amount of flavor immediately after creaming.

Lindsay (1967) suggests 0.5 to 2.0 ppm for synthetic flavor formulations of cultured butter, sour cream, cottage cheese, and buttermilk. These values are probably also the desired values of diacetyl in most cultured dairy products. Increase of diacetyl by addition of citrate or pyruvate to a starter culture may not always be advantageous. Butter with high diacetyl content seems to be more susceptible to oxidation and spoilage (Galesloot and Hassing, 1961).

ALDEHYDES, FLAVOR AND OFF-FLAVORS

We have mentioned briefly the occurrence of acetaldehyde during the biosynthesis leading to diacetyl formation. However, acetaldehyde itself has considerable importance in the formation of flavor in some specific dairy products. Acetaldehyde ($CH_3C \overset{O}{\diagup} - H$, ethanal) has a distinct pungent odor which at suitable low concentrations imparts a characteristic flavor, often described as yoghurt flavor. Yoghurt usually prepared by inoculation of thermophilic lactic acid bacteria — mostly *Str. thermophilus* and *Lb. bulgaricus* (Robinson and Tamime, 1975) — into somewhat thickened milk (16 to 20% dry solids) is incubated for a number of hours (2.5 to 5 or longer) at 45°C. The use of a mixed culture of these two organisms has proved useful and presents another interesting case of associative growth in the group of lactic acid bacteria. The lactobacillus (rods), being somewhat proteolytic, liberates amino acids (especially valine, histidine, and glycine) from the milk proteins (Bautista et al., 1966). Other workers have supported these observations by incubating *Lb. bulgaricus* and *Str. thermophilus* separately in milk and analyzing their respective free amino acids by an amino acid analyzer (Miller and Kandler, 1966). While the *Lb. bulgaricus* culture had a total free amino acid concentration of 963 mg/lt, *Str. thermophilus* yielded only 115 mg/liter. These amino acids are utilized by the rapidly multiplying streptococcus which initiates rapid acidification of the product. Lipolytic activity and the formation of volatile fatty acids in yoghurt have also been attributed to *Lb. bulgaricus* (Turcić et al., 1969). Towards the last period of the incubation schedule, the more acid resistant rods also contribute substantially to the souring, which has to be stopped by rapid cooling in order to provide the desired acidity (0.85 to 0.90%).

The characteristic flavor of yoghurt is formed by acetaldehyde, although there is still some controversy with regard to its producer. Some attribute acetaldehyde formation to *Str. thermophilus* in addition to acetoin and diacetyl (Choisy, 1974). This may be supported by the fact that excellent yoghurt may also be manufactured by the exclusive use of *Str. thermophilus* (Kupsch, 1972). On the other hand, Bottazi and Vescovo (1969) claim that most of the acetaldehyde is produced by the lactobacillus although considerable strain variation could be observed (Table XIII).

However, it is essential that in employing a mixed culture as yoghurt starter the two types of bacteria should be well balanced, a ratio of 1:1 probably being best. Although the cocci propagate much faster, by the time high acidity is reached the acid resistant rods increase in number while the acid sensitive streptococci are inhibited, and the 1:1 ratio is reestablished. Hamdan and co-workers (1971) examined the acetaldehyde-producing potential of these cultures and reported a maximum of 22 to 26 ppm at the fifth hour of incubation. Mixed cultures had higher values than single fermentation (Fig. 17).

Reliable data on the acetaldehyde content of commercial yoghurt preparations are rare. Formisano et al. (1974) gave the values of 4 to 17.5 ppm for acetaldehyde in yoghurt, the higher

TABLE XIII
CARBONYL COMPOUNDS PRODUCED BY THERMOPHILIC LACTOBACILLI

Group	ppm Acetaldehyde	Acetone	Yoghurt Flavor
A	1.43	3.24	slight, not typical
B	3.88	2.44	clean, not very full
C	8.50	3.00	very good and strong

SOURCE: From Bottazi and Vescovo (1969).
NOTE: Only traces of acetoin were detected.

values in examples prepared by slow cooling after the incubation at 42.5°C. Other cultures also seem to be able to contribute to this flavor by varying degrees. Keenan et al. (1966) have shown that even the standard acid producers (*Str. lactis* and *Str. cremoris*) will produce some acetaldehyde (5 to 8 ppm and less according to strains, incubated at 21°C, peak values 10 to 12 hrs). *Str. diacetilactis* also seems to be outstanding here, acetaldehyde val-

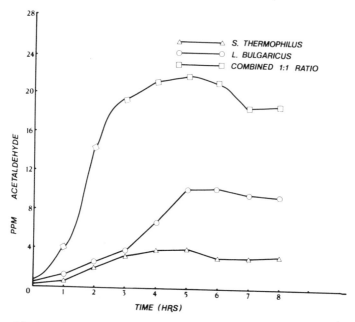

Figure 17. Acetaldehyde Production by a Single Strain of *Str. thermophilus, Lb. bulgaricus* and a 1:1 Mixture of Both. After Hamdan et al. (1971).

ues increasing steadily to 11 ppm within 24 hours of incubation. Hence, acetaldehyde will presumably take part in flavor formation in nonyoghurt cultured milk products too. Indeed, lactic cultures have been described to cause a flavor defect known as "green" or "yoghurtlike." This off-flavor was ascribed to undesirably high acetaldehyde values in cultured buttermilk and sour cream. Several authors have expressed the view (Lindsay et al., 1965; Sandine et al., 1972) that flavor of such products is affected both by diacetyl and acetaldehyde, the absolute quantities determining the intensity, the ratio (diacetyl/acetaldehyde) the flavor balance (Table XIV).

Desirable flavor was found at a ratio of 4.4 to 3.2, lower values indicating green or yoghurtlike flavor, higher values (above 5.5:1) yielding a harsh, unbalanced flavor. Since many cultures show lower acetaldehyde values after peak concentrations, it may be assumed that acetaldehyde is utilized after additional incubation. *Lc. cremoris* seems to be an efficient acetaldehyde scavenger (Keenan et al., 1966), decreasing the acetaldehyde

TABLE XIV

RATIO OF DIACETYL TO ACETALDEHYDE IN MIXED-STRAIN BUTTER CULTURES AND ITS RELATION TO FLAVOR AND AROMA

Culture	pH	Diacetyl (A)	Acetaldehyde (B)	Ratio A:B	Flavor and Aroma
		—(ppm)*—			
1	4.70	7.80	0.60	13.0:1	Harsh
2	4.65	4.44	0.44	10.0:1	Harsh
3	4.60	1.01	0.13	7.8:1	Not full
4	4.60	3.33	0.45	7.4:1	Harsh
5	4.60	1.60	0.25	6.4:1	Not full
6	4.80	3.66	0.66	5.5:1	Harsh
7	4.60	1.65	0.35	4.7:1	Good
8	4.55	2.30	0.52	4.4:1	Good
9	4.50	2.44	0.59	4.1:1	Good
10	4.60	1.50	0.42	3.6:1	Good
11	4.65	1.20	0.38	3.2:1	Good
12	4.80	2.40	2.20	1.1:1	Green
13	4.70	0.96	1.20	0.8:1	Green
14	4.50	2.32	3.88	0.6:1	Green
15	4.80	0.80	1.65	0.5:1	Green
16	4.70	1.00	2.30	0.4:1	Green

SOURCE: From Lindsay et al. (1965).
* Average of duplicate analyses.

concentration in a mixed strain butter culture from 8.5 ppm to 2.63 ppm within 24 hours. Overgrowth of *Str. diacetilactis*, especially in citrated cultures, would aggrevate the situation due to increasing acetaldehyde production (Vẹdamuthu et al., 1964).

A synthetic flavor for cultured butter formulated by Lindsay et al. (1967) suggests an acetaldehyde concentration of 0.2 mg/kg, with a diacetyl content of 0.5 to 2.0 ppm.

A variety of fermented dairy products containing acetaldehyde in addition to other flavor components are marketed in many countries. Since no strict control is exerted in their manufacture, little is known about their composition. Recently Baroudi and Collins (1976) described the composition of *laban* (a Lebanese product also in use in other Mediterranean countries) produced under controlled conditions employing the original Lebanese starter. The following flavoring ingredients were found — average titrable acidity 1.05 percent, pH = 4.2; ethanol 1.25 percent; acetaldehyde 4.2 ppm; acetoin + diacetyl 34 ppm. Organoleptically it had a tart taste, slightly yeasty but with no detectable diacetyl aroma. The following organisms were isolated from the original starter: *Str. thermophilus, Lb. acidophilus, Lb. lactis;* the yeasts isolated were *Saccharomyces cerevisiae* and *Kluyveromyces fragilis.* Interestingly, the formation of acetaldehyde was ascribed to the latter organism and not to the lactic acid bacteria. Although yeasts play quite an important role in certain fermented milk products, e.g. *kefir, koumiss,* little is known about the nature of their flavor components. The use of yeasts for the promotion of flavor in cultured milk similar to yoghurt has been suggested by Kuwabara (1970). By introducing *Kloeckera africana* into the milk, little alcohol but appreciable amounts of aromatic substances were formed. This was followed by sterilization and the introduction of a lactic starter. Such a preliminary fermentation with yeast improved the flavor and consistency of the product. However, no description of the aromatic substances is given.

Grudzinskaya (1968) studies the microflora of the kumiss (mare or cowmilk) fermentation. Although *Lb. casei, Lb. bulgaricus,* and acetic acid bacteria were isolated, the fermentation of lactose by yeasts and the formation of alcohol (about 2.5%) seems to be the more important fermentation, imparting the characteristic effervescent character of the product. Also in leben, a fermented milk produced on farms in Iraq and other eastern countries (1.3% titrable acidity, 0.076% ethanol, 309 mg acetoin + diacetyl/kg), lactose-fermenting yeasts are important in the formation of the characteristic flavor (Abo-Elnaga et al., 1977). So far, little research has been directed toward the elucidation of flavor compounds in such cultured milk production. An attempt to study the chemical composition of the flavor of yeast fermented milk was carried out in a Japanese skim milk beverage fermented with *Candida pseudotropicalis, Torulopsis sphaerica,* and a mixture of *Str. lactis* and *Lb. bulgaricus* (Nakanishi and Arai, 1973). The following volatile carbonyl compounds were detected — acetaldehyde 3.5 to 44 ppm, acetone 0.8 to 16 ppm, and 2-butanone 0.2 to 5.7 ppm. The lower values were obtained when the beverage was fermented in the absence of yeasts.

Acidophilus milk, which is widely advocated for therapeutic use in certain gastric disorders, differs from yoghurt in its less buttery and more astringent flavor. These differences are probably due to the different metabolism of *Lb. acidophilus,* although the exact nature of the flavoring components found in the somewhat unpleasant preparation of acidophilus milk has not yet been worked out (Davis, 1963). More recently, it was pointed out that *Lb. acidophilus* produces large amounts of 2-heptanone when cultivated in skim milk. This may be related to the unpleasant acidophilus milk flavor (Yu and Nakanishi, 1975). However, the production of sweet acidophilus milk seems to be possible at this time (Speck, 1975).

THE MALTY DEFECT

In addition to the green flavor defect mentioned earlier, another off-flavor is known to occur in cultured dairy products. This relates to the malty flavor — described as burnt, caramel, or grapenut flavor — of fermented milk. This defect has been known for a long time and was ascribed to the appearance of a lactic variant, *Str. lactis* var. *maltigenes,* or the recently described new species *Lb. maltaromaticus* (Miller et al., 1974). The nature of this off-flavor has been studied more thoroughly only in more recent years. GLC analysis of head space of "malty" lactic cultures revealed that the principal component of the malty defect was 3-methyl butanal; 2-methyl propanal, 2-methyl propanol, and 3-methyl butanol were also involved (MacLeod and Morgan, 1958). Since the odor of 2-methyl propanal is similar to that of 3-methyl butanal, the role of this compound may be considered as largely additive. On the other hand, 2-methyl propanol and 3-methyl butanol, the aroma of which are related to their corresponding aldehydes, are less harsh and seem to modify the acid character of the aldehydes in the aroma (Morgan et al., 1966).

Strains of *Str. lactis* are known to possess an active transaminase system which mediates the transfer of the amino group of leucine to α-keto-glutaric acid, resulting in the formation of α-keto-isocaproic and glutamic acid:

$$(CH_3)_2CH\ CH_2\underset{\underset{NH_2}{|}}{C}HCOOH + COOH(CH_2)_2\ \overset{O}{\overset{\|}{C}}\!\!-\!\!COOH \xrightarrow[\text{phosphate}]{\text{pyridoxal}} \Longleftrightarrow$$

$$(CH_3)_2CHCH_2\overset{O}{\overset{\|}{C}}\!\!-\!\!COOH + COOH(CH_2)_2\underset{\underset{NH_2}{|}}{C}HCOOH$$

α-KETO-ISOCAPROIC ACID

These experiments were carried out with resting cells of the corresponding strains.

The maltigenic variant is probably the only strain that possesses an additional enzyme that decarboxylates this keto acid to 3-methyl butanal:

$$(CH_3)_2\ CHCH_2\overset{O}{\overset{\diagup\!\!\!/}{C}}\bullet COOH \xrightarrow[\text{pyrophosphate}]{\text{thiamine}} (CH_3)_2\ CHCH_2\overset{O}{\overset{\diagup\!\!\!/}{C}}\!\!-\!\!H + CO_2$$

but may be reduced to the corresponding alcohol (after Morgan, 1976).

$$R\bullet\overset{O}{\overset{\|}{C}}\!\!-\!\!H \xrightarrow{\text{NADH}_2} R\bullet CH_2OH$$

Other amino acids may yield, in an analogous way, the following aldehydes:

isoleucine   → 2-methyl butanal
valine   → 2-methyl propanal
methionine   → 3-methyl thiopropanal (cooked cabbage odor)
phenylalanine   → phenylacetaldehyde (flowerlike aroma)

The simultaneous detection of ethanol, 2-methyl propanol, and 3-methyl propanol in the presence of the corresponding aldehydes is believed to be rather conclusive evidence for the presence of a yeastlike alcohol dehydrogenase(s) in this organism. In a more recent publication, the Corvallis group (Sheldon et al., 1971) presented additional information on the chemical nature of the malty defect. Employing GLC and mass spectrometric analysis, they also revealed phenylacetaldehyde and phenethyl alcohol in ether extracts of steam distillates of malty cultures, which did not appear in control distillates of heated milk.

As may be seen from Table XV, the threshold value of the aldehydes is much inferior to that of the corresponding alcohol. It appears that strong maltiness is correlated with a decrease in the alcohol/aldehyde ratio, especially of 3-methyl butanol/3-methyl butanal.

In practice, a starter containing 3 to 5 percent of the malty flavor producing bacteria will always develop a malty defect. However, it was shown that in presence of *Lc. cremoris* the reduction of 3-methyl butanal will take place, eventually preventing maltiness. A method for the detection of malty flavor producing streptococci has been reported (Langeveld, 1975).

FLAVOR FORMATION IN RIPENED CHEESES

Most of the biosynthetic processes mediated by bacteria described in the chapter dealing with cultured dairy products may also take place in the manufacture of ripened cheeses. However, the aging of cheeses is a much more complex process that comprises a number of factors that do not appear or at least do not seem to be of any significance in fermented milk products with comparatively short shelf lives.

TABLE XV
FLAVOR THRESHOLD FOR SOME VOLATILE COMPOUNDS
IN MALTY CULTURES

Compound	ppm
2-methyl propanal	0.10
2-methyl propanol	5.00
2-methyl butanal	0.13
2-methyl butanol	6.25
3-methyl butanal	0.06
3-methyl butanol	3.20
Phenylacetaldehyde	0.02
Phenethyl alcohol	0.07

Ripened cheeses may be classified as — (1) hard grating cheese (the Parmesan types); (2) hard cheese (Swiss, cheddar, Gouda, provolone, etc.); (3) semihard cheese (Roquefort®, blue, Limburger, etc.); (4) soft cheese (Camembert, Liederkranz). Practically hundreds of cheese varieties are known throughout the world, providing a broad spectrum of flavors for the human palate (Sanders, 1953). Origin of the milk, fat and water content, amount of salt incorporated at certain stages in their manufacture, starter cultures, curd formation and compaction into final shape, ripening cultures employed and aging conditions, all affect the quality and, of course, the flavor profile of the final product. Even though there are so many varieties, all have certain common features in the processes that lead to flavor formation, at least with respect to their broad classification.

Although the microbial activities are usually considered to be of primary importance in the formation of the flavor pattern, it has to be kept in mind that a variety of processes of physical-chemical nature will affect the manufacture of cheeses and will ultimately also contribute to the final flavor profile. Experiments described below in which cheese production could be carried out under aseptic conditions have indeed yielded considerable information as to the nonmicrobial changes that cheeses undergo during processing and aging. The microbial part in cheese making may also comprise a number of categories:

1. The microflora derived from the milk intended for cheese

making: raw or pasteurized milk; sanitary conditions prevailing during and after milking.

2. Nature of starters used before or with enzyme clotting of the curd (rennetting).
3. Appearance and reproduction of other bacterial cultures, the "adventitious bacteria," usually derived from the environment during processing.
4. The biochemical activities of these microorganisms, during processing and aging, comprising activities due to autolytic processes taking place on prolonged incubation.
5. Special cultures introduced for cheese curing.

GENERAL APPROACH TO CHEESE FLAVOR ANALYSIS

A water suspension of cheese with characteristic flavor may be separated into taste and aroma fractions by vacuum distillation (Harper, 1959). The residue will have a cheeselike taste but will definitely lack in aroma. On the other hand, the distillate will have a pleasing cheesy aroma but will be lacking in cheese taste. Those chemicals, which would remain in the residue after vacuum distillation and could contribute to the "base" or taste of such a cheese, would usually include lactic acid, amino acids, nonvolatile amines, salt, and various fragments of the proteins and fats. On the other hand, the compounds that would contribute to the aroma portion of the cheese flavor would include such groups as amines; fatty acids; aldehydes; ketones; alcohols; esters; and volatile sulfur compounds, such as H_2S and mercaptanes.

Analysis of the various components that constitute the flavor, taste, and aroma fractions of the cheese is still not sufficient in determining which of these compounds actually create the flavor sensation. This, of course, is extremely important to the flavor chemist who attempts not only to understand the nature of flavor in cheese but also wishes ultimately to imitate the taste and aroma of good cheeses by synthetic means. To relate the objective chemical analysis to subjective flavor evaluation is another major problem. Even between experts there may be a lack of agreement as to the degrees of perfection of a certain flavor and its independence of other flavor components. In order to establish the relationship of a specific chemical compound with the characteristic flavor of the cheese, the following criteria have been used — (1) the presence of the specific compound, (2) an increase in the concentration of the specific compound as the ripening progresses, (3) variations in the rate of formation of the constituent in raw and pasteurized milk made cheese, (4) association between the concentration of the compound and the intensity of the characteristic flavor of the cheese, (5) reproduction or enhancement of the characteristic flavor when the compound is added to cheese or to the taste base. Of these, the last two may be the most valuable criteria (Harper, 1959). A number of difficulties must also be considered here. It is possible that a certain flavor compound may be produced simultaneously with another flavor constituent. It may serve as a precursor to another even more powerful flavor determinant. On the other hand, the failure of a certain compound to reproduce or enhance the flavor of a product does not mean that it does not actually contribute to the flavor, since it is almost impossible to add back the specific compound in the exact chemical and physical state in which it existed originally or to have the accompanying constituents available that might be essential for the added compound to exert its influence on the flavor pattern of the modified product. Hence, it is difficult, if not impossible, to predict the effect that the addition of one flavor component will have on a complex mixture of flavor chemicals. Blending, enhancement, or masking of flavors still remain unpredictable features of flavor chemistry. Nevertheless, the qualitative and quantitative description of the chemical compounds that can be found in cheese as a function of its ripening process remains the most powerful means in the study of the flavor chemistry of ripened cheeses. Even the presence of compounds below their flavor threshold concentrations may be of importance due to possible synergistic or enhancing properties, still little understood. Most authorities believe that typical cheese flavor results from blending of a variety of specific individual substances in proper proportions — the "Component Balance" theory of flavor (Mulder, 1952; Kosikowski and Mocquot, 1958). A complete understanding of cheese flavor would have broad commercial significance in permitting control of uniformity of flavor, avoiding economic losses and increasing sales potential.

At this point, knowledge of the basic steps of cheese technology would be advantageous. Since many excellent reviews and books are available on this subject (Robertson, 1966; Davis, 1976; Kosikowski, 1977), the reader is referred to them. Some of the critical steps concerning flavor formation will be briefly mentioned (Fig. 18). A comprehensive treatment of the biochemistry of cheese ripening is also available (Schormüller, 1968).

THE MICROBIAL LOAD OF MILK

Total counts of microorganisms in milk arriving at the dairy reflect to a great extent the sanitary conditions prevailing during and after milking (Thomas et al., 1971). Counts may vary from 10^4 to 10^6 per ml. Many countries have made pasteurization compulsory in order to protect the consumer from sanitary hazards. However, these regulations do not always affect milk

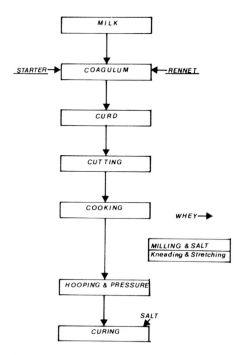

Figure 18. Major Steps in Cheese Production.

designated for cheese production with long ripening periods (Cheese and Cheese Products, Definitions and Standards, 1971). Pasteurization of milk will reduce the number of many heat sensitive lactic organisms (not *Str. fecalis*) as well as many of the undesirable gas-forming organisms and a concurrent inactivation of some of the milk enzymes — alkaline phosphatase inactivation provides a reliable index of the adequacy of heating (Wong, 1974). However, the greater part of the microbial load survives pasteurization (usually 15 seconds at 73°C). Numbers of 10^3 to 10^5 per ml may be found in literature (Franklin and Sharpe, 1963). A number of authors have compared the suitability of pasteurized and raw milk for cheddar cheese manufacture and have reached the conclusion that typical cheddar cheese flavor may be obtained from heated milk provided the bacterial count was neither abnormally high nor too low. The positive effect of the initial bacterial load on flavor formation, especially in strong flavored cheddar cheese, has been pointed out (Price and Call, 1969; Ohren and Tuckey, 1969). On the other hand, heated milk is generally not considered satisfactory for the manufacture of Swiss cheeses. Whether this is due to the tendency of overgrowth of anaerobic bacteria, which cause defective eye formation, or other reasons has not been established with certainty. Discussion follows later in this chapter. Emmental(er) cheese of good quality has, however, been made from heated milk. Mold ripened cheeses have been produced both with raw and heated milk.

FLAVOR DEVELOPMENT IN CHEDDAR CHEESE

The flavor of cheddar cheese is associated with a pleasant, slightly sweet, aromatic, walnut sensation, without any outstanding single note. In aged cheese, a biting quality, which is neither coarse nor unpleasant, gives sharpness to the cheese (Kosikowski, 1957).

Milk Fat Hydrolysis

The major flavor constituents of cheeses are the degradation products of fat and protein of the milk. Milk fat is perhaps the most important ingredient; cheese made from skim milk does not develop a typical cheese flavor (Ohren and Tuckey, 1969).

Only cheese with 50 percent fat (dry basis) or more develops a typical flavor. Typical cheddar cheese flavor was found to be related to a balance of *free fatty acids* (FFAs) and acetic acid. Experimental lots of cheese, which had the finest flavor and highest flavor score, had a concentration of FFA plus acetate of 12 to 28 μmoles/g of cheese solids after 18 days of ripening or a ratio FFA/acetate between 0.55 and 1.0. All even numbered carbon fatty acids (C_4 to C_{18}) could be demonstrated from the first day of manufacture. Cheese criticized as having a rancid, fruity (see following text), and fermented off-flavor had 2 to 3 times higher concentrations of C_{10}, C_{12}, and C_{14} FFA than cheeses of high quality. Cheeses made from milk with lower fat content showed a lower FFA concentration, while the acetic acid concentration increased so that the FFA/acetate ratio became undesirable. However, the importance of the FFA/acetate ratio to cheese flavor has been recently questioned (Reiter and Sharpe, 1971).

Kristoffersen and Gould (1960) adopted a somewhat different view in evaluating the flavor of ripened cheddar cheese. According to these authors, H_2S produced during the aging process plays an important role in flavor formation. The ratio of FFA (including acetic acid) to H_2S determines the final quality of the cheese (Table XVI).

Best flavored, balanced cheeses had a FFA/H_2S ratio (μmoles/g/μmoles/100 g) of 12 to 17. Lawrence (1963a), employing a modified entrainment method for the determination of microquantities of H_2S, found levels of 45 to 60 μg of hydrogen sulfide per 100 g in New Zealand cheddar ripened 3 months. These values would remain constant upon additional aging. The possibility that additional products are formed between H_2S and various carbonyl, carboxyl, and ester groupings, which may affect flavor, has also been advanced. The flavor threshold for H_2S is around $1:10^{10}$ parts. In addition, the formation of active SH groups during cheese ripening is considered to be of importance in the development of the desirable organoleptic qualities of cheeses (Kristoffersen, 1967).

The origin of cheese fatty acids has received considerable attention. Acetic acid seems to have a variety of sources. While almost no lactose remains in the curd, a number of intermediates

TABLE XVI
FFA CONTENT OF CHEDDAR RIPENED 18 MONTHS

FFA	mole/g Cheese
Acetic	2-26
Propionic	0.1-1.9
Butyric	0.1-1.7
Higher FAs	2.1-15.5

derived from the sugar or citric acid may give rise to acetate (Reiter and Møller-Madsen, 1963). Most of the higher carbon fatty acids are derived from milk fat through bacterial lipolysis. Although milk also contains a number of lipases, the pH of cheddar cheese (5.2 to 5.4) is not optimal for their activity. Since milk fat is comparatively rich in esterified butyric acid, it is not surprising that a considerable percentage of FFA occurring in ripened cheese is butyric acid. Bills and Day (1964) examined the FFA content of a large number of cheddar cheeses and found the content of butyric acid to vary between 90 and 900 mg/kg cheese. The occurrence of about twice as much butyric acid than the esterified form in milk fat seems to be indicative of the specific action of these lipases on milk fat. Both lipases from animal sources frequently added prior to casein coagulation, as well as milk lipases, are known to selectively hydrolyze the 1-position of milk triglycerides, which is predominantly esterified by butyric acid (Clement et al., 1962). According to Forss and Patton (1966), acetic, butyric, and caproic acids appear to be indispensable for the formation of good flavored cheddar. They suggest an optimal ratio of 8 : 1 : 0.3 (or 900 acetic/110 butyric/35 caproic acid, ppm). Incubation of milk with *Geotrichum candidum* prior to cheese making increased the concentration of C_6 and higher FAs (Irvine et al., 1954).

Other fatty acids up to C_{18}, saturated and unsaturated, also appear in cheddar cheese. Quantitatively, the 14:0, 16:0, 18:0, 18:1 FFA are predominating. Although detailed studies on the microbial lipase activity on milk fat are not abundant, it seems likely that microbial systems take an important part in the hydrolysis of unsaturated fatty acids from milk fat (Khan et al., 1966). *Achromobacter lipolyticum* seems to be particularly active in the liberation of oleic acid (18:1) from milk fat (Khan et al., 1967). Franklin and Sharpe (1963) investigated the nonstarter bacterial population of cheddar cheeses. After 3 months ripening, values of 10^7 to 10^8 per gram cheese could be found. A correlation between higher bacterial counts and flavor score was established with a difference of 4 to 18 percent over the score for cheeses with poorer bacterial counts. In addition to *Lb. lactis* and *Lb. plantarum* mentioned below, *Lb. brevis* and *Pediococcus* spp. were also found. *Lb. casei* predominated especially when more severe pasteurization regimes (17 seconds at 76.7°C) were employed, indicating some higher heat resistance of these bacteria. Of the nonstarter, nonlactic bacteria examined in cheeses ripened for 3 months, 13 percent of the isolates were able to hydrolyze butterfat (about one-half were sporeformers, the rest were gram-positive cocci). We shall return to microbial lipases in connection with mold ripening of cheeses (see the section Flavor Formation in Mold Ripened Cheeses later in this chapter). The question naturally arises concerning the role played by the lactic starter organisms in milk fat hydrolysis. Usually *Str. cremoris* or a mixed culture of *Str. lactis, Str. cremoris,* and *Leuconostoc cremoris* (occasionally also thermoduric lactics) are employed, allowing the acidity to reach 0.6 to 0.75 percent. Stadhouders and Veringa (1973) claim that starter culture bacteria show only limited capacity to hydrolyze milk triglycerides, although they are capable of hydrolyzing mono- and diglycerides fairly rapidly. Hence, starter organisms may cooperate with milk lipases and the lipases of gram-negative bacteria in the hydrolysis of milk fat. These observations were substantiated by the work of Chandan et al. (1969) who studied the lipolytic activity on tributyrin by lactic acid bacteria grown in fermentors. Lipase activity of supernatants was insignificant. Cellular lipase activity of these organisms (FA liberated, μmole/mg DNA) ranged from 0.7 to 32.8, *Lc. dextranicum* being the most active lipolytic organism. Weak lipolytic activity versus tributyrin with lactic acid bacteria was also reported by Oterholm et al. (1968). A method for the detection of weak lipolysis by dairy lactics on double layered agar plates is suggested by Umemoto (1969).

On the other hand, lactic acid bacteria seem to be quite active with regard to the formation of volatile fatty acids (VFA). The mechanism of formation of VFA involves a deamination or decarboxylation (or both) of certain amino acids. Nakae and Elliot (1965) studied the formation of VFA with strains of *Str. lactis, Str. diacetilactis*, and some lactobacilli and could demonstrate the production of acetic acid from serine, alanine, and glycine; propionic acid from threonine; isobutyric acid from valine; isovaleric acid from leucine; and 2-methyl butyric acid from isoleucine. This contribution of lactic acid bacteria and possibly also other bacteria requires the presence of free amino acids, which are derived from the breakdown of milk proteins during cheese ripening.

Milk Protein Degradation

Freshly made cheddar cheese has a firm, elastic, "curdy" body with a mildly acid flavor and slight aroma derived mostly through the fermentative activities of starter organisms. Firmness and elasticity become reduced during the early period of ripening (2 to 4 weeks). These changes in body texture are accompanied by a conversion of part of the protein to water soluble compounds, peptones, and free amino acids. As ripening continues, the percentage of free amino acids increases. Concurrently, an increase in the ammonia concentration takes place. After several months of ripening, the ammonia content may reach over 3 percent — that of free amino acids over 12 percent — of the total nitrogen compounds of the cheese (Foster et al., 1957).

That the abundant population of raw milk is very effective in milk protein degradation may be derived from the fact that cheeses prepared from raw milk had much higher percentages of free amino acids than those prepared under similar conditions with pasteurized milk. Higher values were recorded for aspartic acid, leucin, methionine, glutamine (but not glutamic acid). These differences were held responsible for the somewhat sharper flavor of raw milk cheddar (Kosikowski, 1951). Harper (1959) found a 10 to 20 percent increase in the free amino acid content (including glutamic acid) of raw milk cheddar. Glutamic acid increased from 5 to 10 mg/10 g cheese solids of young (1 month) flavorless cheese to 50 to 60 mg/10 g cheese solids in a ripened (8 months) cheddar with definite flavor. However, it is now generally agreed upon that the amino acids, amines, and peptides fraction do not per se impart cheese flavor but rather appear to provide the "brothy" background on which the typical flavor becomes superimposed (Day, 1967).

The source(s) of the enzyme(s) responsible for protein breakdown in cheddar are not known with certainty. It is generally accepted that rennet is mediating the initial steps in which large protein molecules are split to smaller proteoses and peptones, while the hydrolysis to free amino acids and liberation of ammonia and amines are attributed to microorganisms (Silverman and Kosikowski, 1956). What role, if any, do the lactic starter organisms play in these hydrolytic reactions? Some indication of the proteolytic potential of lactic starter organisms may be obtained from the work of Zalashko et al. (1970). Tyrosine and tryptophan liberation (Folin and Ciocalteau reagent) was determined after prolonged incubation (72 hours) in a milk medium (Table XVII).

Following the distribution of starter organisms during cheddar ripening, it becomes clear that the number, which initially may reach many billions per gram cheese, is declining rapidly, reaching only several millions and less after one month, *Str. lactis* apparently holding out somewhat longer. During this period, a new population takes over, consisting chiefly of lactobacilli (*Lb. casei* and *Lb. plantarum*) not originating from the starter bacteria. These organisms are believed to take part in the hydrolytic processes, especially after the first weeks of ripening. A peptide

TABLE XVII
PROTEOLYSIS BY LACTIC STARTER ORGANISMS

Organism	*Tyrosine + Tryptophan* *mg%*
Str. thermophilus	0.37-1.84
Str. paracitrovorus (Lc. dextranicum)	0.5
Str. lactis	9.0
Str. diacetylactis	8.75-10.0
Lb. casei	9.2
Lb. bulgaricus	11.7
Lb. plantarum	12.8
Lb. helveticus	16.8
Lb. acidophilus	18.6

consisting of sixteen amino acid residues isolated from extracts of *Str. lactis* was shown to be stimulatory to *Lb. casei* (Branen and Keenan, 1970). Attempts have been made to introduce lactobacillus cultures to milk for cheese (Johns and Cole, 1959). These attempts were not always successful since cheeses prepared in this way developed a harsh or fermented flavor. Surface bound extracellular proteases, as well as intracellular caseolytic enzymes in lactic bacteria, have been described recently (Gripon et al., 1977; Schmidt et al., 1977). The proteolytic activity of lactic organisms during cheese maturation cannot, therefore, be ignored.

Carbonyl Compounds

The effort to find a "specific cheddar flavor compound" has directed the attention of many flavor chemists toward compounds with a volatile nature that occur only at minute concentrations. More sophisticated analytical methods also made it possible to detect and evaluate compounds at ppm levels.

Much of our knowledge on the carbonyl compounds occurring in cheddar cheese we owe to Day and his collaborators (1960). These were isolated from distillates of cheese slurries and studied as 2,4-DNP hydrazones after separation and purification by column partition chromatography (Table XVIII).

Although many of the methyl ketones occur at high concentrations in ripened mold cheeses (discussion later in this chapter), none of the cheddars examined had mold growth. The significance of these compounds has been questioned. Some authors believe them to be artifacts resulting from the degradation of β-keto esters, which are normal constituents of milk fat (Lawrence, 1963b; Ven et al., 1963). However, Scarpellino and Kosikowski (1962), who studied 10-months-old cheddar, believe that the odd numbered methyl ketones are derived from the even numbered fatty acids from milk fat, after conversion to β-keto acids and their decarboxylation to produce the methyl ketones. The even numbered methyl ethyl ketone (butanone) would require another mechanism, unless a C_5 carboxylic acid resulting from the deamination of a 5 carbon amino acid had taken place. No such acid was so far discovered in cheddar cheese. Another source has been suggested for this methyl

Flavor Microbiology

TABLE XVIII
CARBONYL COMPOUNDS IN CHEDDAR CHEESE

Compound	Approximate Concentration mg/kg Cheese
2-Tridecanone	2.10
2-Undecanone	1.43
2-Nonanone	0.68
2-Heptanone	0.82
2-Pentanone	0.37
3-Methyl butanal	trace
Butanone	12.50
Acetone	8.50
Ethanal	7.50
Propanal	2.60
Methanal (Formaldehyde)	1.0
3-Methyl thiopropanal	0.1
2,3, Butanedione	0.68

SOURCE: After Day et al. (1960).

ketone. Since acetoin has been shown to be reduced to 2,3-butanediol by a variety of microorganisms, it is possible that this can be further transformed into methyl ethyl ketone under the reducing conditions that prevail during ripening of the cheese:

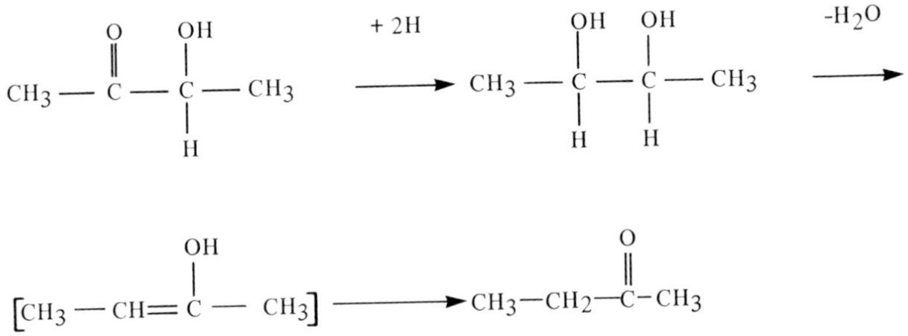

METHYL ETHYL KETONE

Walker and Keen (1974) examined the concentrations of odd numbered (C_3 to C_{15}) methyl ketones in cheeses and found that their values increased very slowly during a one year's ripening period, reaching values far below those that could be expected from the complete breakdown of esterified β-keto acid precursors in the fat. These results would support the earlier view (Lawrence, 1963b) that methyl ketones are formed slowly in cheddar fat because of the inherent instability of the esterified β-keto acid precursors, even at low ripening temperatures. Their formation may be aided by limited lipolytic activity (milk lipases, bacteria). The comparatively high *flavor threshold value* (FTV) of methyl ketones explain why 2-heptanone and to a certain extent also 2-pentanone and 2-nonanone may have a minor but direct contribution to cheddar flavor (Table XIX).

TABLE XIX
FLAVOR THRESHOLD VALUES FOR METHYL KETONES IN CHEDDAR

Compound	FTV, ppm	Found in Cheese, ppm
C4	25-80	3-12
C5	0.5-8.4	0.4-0.8
C7	0.25-0.7	0.8-1.9
C9	0.25-3.5	0.5-0.7

Source: After Walker and Keen (1974).

Much has still to be learned about the biochemistry of ketone formation in cheeses. Many of the reactions postulated to be mediated by microbes should be demonstrated in bacteriologically controlled systems.

Some aldehydes have also been identified in cheddar cheeses. Methanal (formaldehyde), ethanal, propanal, 3-methyl butanal, and 3-methyl thiopropanal (methional) have been identified (Day and Libbey, 1964). These compounds may result from the Strecker degradation of amino acids (oxidative deamination and decarboxylation) or the transamination and decarboxylation of the corresponding amino acids by cheese bacteria (MacLeod and Morgan, 1958), e.g. methionine + pyruvate $\rightarrow$ methional + alanine (Keeney and Day, 1957). Methional is very unstable and may account for the occurrence of methyl mercaptan and dimethyl sulfide in cheddar cheese (Patton et al., 1958). Among the alcohols, ethanol and 2-butanol appear in cheddar. Apparently, some of the bacteria associated with the ripening process possess alcohol dehydrogenases that reduce the ketone and aldehydes formed in cheese to the corresponding alcohols (Scarpellino and Kosikowski, 1962). Microorganisms related to the formation of 2-butanone and 2-butanol have been recently identified (Keen et al., 1974). Three steps were involved:

1. The formation of 2,3-butylene glycol by the reduction of acetoin may be carried out by certain starter organisms, especially *Str. diacetilactis RAC1* as well as *Pediococcus cerevisiae* 12.
2. 2,3-butylene glycol may be transformed into 2-butanone (methyl ethyl ketone) by *Lb. plantarum* 20.
3. 2-butanone may be reduced to 2-butanol by *Lb. brevis* 24.

All these reactions were carried out with resting cells of the respective cultures and substrates and may be regarded as strong evidence in favor of the involvement of nonstarter bacteria in the formation of ketones and their corresponding alcohols. However, their role in flavor buildup may be only secondary. Our knowledge on the contribution of bacteria to these specific reactions in cheese is still meagre. We shall discuss the role of esters in cheese in connection with some defects occurring in certain cases.

Flavor Defects in Cheddar Cheeses

Fruity Flavor

One of the most common off-flavors encountered in cheddar cheese has been named "fruity." This defect may be described as "pineapple," "apple," or "pear-like." It was generally assumed that this off-flavor was connected with the formation of certain esters in cheeses. However, other compounds such as certain aldehydes and ketones, which are known to have fruity odors and which occur in cheese, may also be involved. Some investigators have produced the fruity defect experimentally by using certain starter organisms. Perry (1961) has shown that employing various starter organisms as single strain cultures for cheddar cheese manufacture, *Str. lactis* cultures consistently produced cheeses with a flavor described as "fruity" or "dirty." Other cultures *(Str. cremoris)* produced cheese with normal flavor. The term "lactis" flavor has been coined in order to describe the defect caused by this organism. These observations were confirmed by Vedamuthu et al. (1964) who demonstrated a strong fruity flavor produced by a mixed culture of *Str. lactis* and *Str. diacetilactis,* while cremoris cultures gave normal cheeses.

The volatile constituents of cheddar cheese with fruity flavor were compared with the volatile constituents of normal cheese produced under identical conditions with the exception of the starter organism employed (Bills et al., 1965). Sensory evaluation of the components isolated from fruity cheese by a molecular distillation technique and separated by gas liquid chromatography have shown that the most important constituents of fruity flavor were ethyl butyrate and ethyl hexanoate (Table XX).

TABLE XX

RELATIVE AMOUNTS OF FRUITY FLAVOR CONSTITUENTS IN CHEDDAR*

	Ethyl Butyrate	*Ethyl Hexanoate*
Fruity	7.84	0.35
Control	1.52	0.05

* Relative peak heights of one pair of experimental cheeses.
Source: Bills et al. (1965).

As can be seen from the example cited in Table XX, the esters present in both cheeses were at much higher concentration in the fruity cheese. Also the levels of ethanol were much higher. These results seem to suggest that excessive production of ethanol may be responsible for the enhanced esterification of some FFA, yielding higher values of the corresponding esters. Since the levels of FFA were comparatively high in twelve cheddar samples examined (865 mg acetic acid, 115 mg butyric acid, 38 mg hexanoic acid, and 41 mg octanoic acid per kg cheese), it was postulated that the level of ethanol may be the limiting factor is ester formation. The importance of ethyl acetate in fruity flavor seems to be secondary only. On the other hand, levels of ethyl octanoate in fruity cheese were not higher than in normal cheddar. This was explained by the fact that octanoic acid is highly lipid soluble whereas the ethanol is practically lipid insoluble. (However, see also McGugan et al., 1975.) If indeed ethanol is primarily responsible for the development of fruity flavor, the microbial production of ethyl alcohol during the ripening process deserves special attention. The production of acetaldehyde by lactic bacteria, especially by *Str. diacetilactis*, has been discussed earlier in connection with the "green" flavor defect of cultured milk products. It is very likely that acetaldehyde acts as a hydrogen acceptor, yielding the required ethanol for the esterification of the free fatty acids. The production of ethyl esters of C_4 and C_6 fatty acids has been studied in greater detail by Hosono et al. (1974) who worked with cell free extracts of lactic acid bacteria and some psychotrophic organisms isolated from milk. They found that the esterase activity of lactic bacteria was much inferior to those of the psychotrophs examined (Table XXI).

Strongest ester-forming activity by pseudomonads has been attributed to the species *Ps. fragi*. Reddy et al. (1968) demonstrated pronounced fruity flavor when this organism was cultivated in sterile homogenized milk. After the typical aroma developed (10 to 12 days at 7 to 21°C), the volatile constituents were examined by GLC and mass spectrometry. Ethyl acetate, ethyl butyrate, ethyl isovalerate, and ethyl hexanoate were identified. Introduction of ethanol (0.2%) markedly enhanced ester production. Ethyl butyrate (0.35 ppm) and ethyl hexanoate (0.5 ppm) were predominating. Fruity flavor formation by *Ps. fragi* may be described by the following formulae (after Morgan, 1976):

I   Fatty Acid Formation:

$$\begin{bmatrix} O-C\overset{O}{\diagup}-R_1 \\ O-C\overset{O}{\diagup}-R_2 \\ O-C\overset{O}{\diagup}-R_3 \end{bmatrix} \xrightarrow{\text{lipase}} R_3-C\overset{O}{\diagup}-OH + R_1-C\overset{O}{\diagup}-OH$$

$$+ \begin{bmatrix} OH \\ O-C\overset{O}{\diagup}-R_2 \\ OH \end{bmatrix}$$

II   Ester Formation:

$$R_3-C\overset{O}{\diagup}-OH + CH_3CH_2OH \xrightarrow{\text{esterase}} R_3C\overset{O}{\diagup}-OCH_2CH_3 + H_2O$$

$$R_3 = -(CH_2)_2CH_3$$
$$-(CH_2)_4CH_3$$

TABLE XXI
ESTER-FORMING ACTIVITY BY SONICATES OF
SOME BACTERIAL CULTURES*

Organism	Ethyl Butyrate	Ethyl Hexanoate
Str. diacetilactis ATCC 15346	0.30	0.15
Str. lactis ML3	0.30	0.04
Str. cremoris TR	0.26	0.23
Lb. casei	0.52	0.01
Pseudomonas No. 50	1.38	3.13
Pseudomonas No. 53	2.06	0.68

SOURCE: After Hosono et al. (1974).
* Activity as units/gr wet cells. 1 unit = 1 μmole ester formed in 3 hours. Data are from shaking cultures.

In spite of the active ester formation by *Ps. fragi*, there are no reports relating fruity flavor in cheddar to this organism. *Ps. fragi* was isolated in a number of cases from fruity milk and cottage cheese. A simple method for the detection of bacteria capable of producing fruity flavor has been recently described (Blais et al., 1978).

Bitter Flavor

A frequent flavor defect of cheddar cheese has been described as the development of a bitter taste during cheddar and other cheese ripening. There seems to be a difference in the sensitivity to this defect among various people; Japanese are reported to be highly sensitive to bitter cheeses (Pelissier and Ribadeau-Dumas, 1976). It is generally accepted that this off-flavor is due to the formation of certain compounds of peptidic nature (Czulak, 1959). A bitter peptide was isolated from nonrennet streptococcal milk cultures and was found to consist of nineteen residues of eight different amino acids: arginine, glutamic acid, glycine, isoleucine, leucine, phenylalanine, valine, and proline (Gordon and Speck, 1965; Zvyaginstev et al., 1972).

There is much less agreement as to the factors that lead to the appearance of this bitter principle. One only knows with certainty that fresh curds are never bitter, so that its formation must be the result of the ripening process. The following factors have been suggested: (1) unsuitable starter organisms, (2) certain nonstarter organisms, (3) low salt content, (4) rennet substitutes,

(5) low cooking temperatures, (6) high heat treatment for cheese milk (Emmons et al., 1962). It is possible that all these factors are related to the proteolytic process involved in the breakdown of a certain casein fraction (probably fraction α_{s1}); caseins practically devoid of this fraction, like in goat milk, will not turn bitter (Pelissier and Ribadeau-Dumas, 1976). Bacterial activity is apparently strongly involved in this process. This may be deduced from the following observations. Breene et al. (1964) found that curds prepared with lactic acid (without starter) always yielded bitter cheeses, while starter curds gave a product with normal taste. Other workers have shown that fast acid formers produce much more bitter cheeses than slow acid producers (Emmons et al., 1962; Lawrence and Pearce, 1972). The involvement of starter cultures in the formation of the bitter principle seems to be quite convincing. Although high rennet concentrations will produce a bitter taste, such concentrations are never employed in cheese manufacture. The question remains whether certain starter organisms themselves produce the bitter principle by a specific hydrolysis of the casein micelle or whether the peptide formed by other proteolytic systems (rennet, nonstarter microflora) is hydrolyzed to nonbitter components by certain starter bacteria as suggested by Czulak (1959). Development of bitterness in lactic acid (nonstarter) curds would be in line with the latter possibility. Another hypothesis would be that proteolysis by rennet results in the formation of a high molecular weight peptide, nonbitter by itself but which serves as substrate for starter bacteria for the liberation of the low MW bitter peptide. In this case, further peptidase activity of starter bacteria would also degrade the bitter peptide. Indeed, prolonged incubation will reduce the bitterness of the cheese. The New Zealand school (Lawrence et al., 1972) emphasizes the predominant importance of starter organisms with regard to the bitter defect of cheese. According to these authors, bitterness will not be appreciable if the correct slow acid forming starter organism, e.g. *Str. cremoris* strain AM_2, is employed. On the other hand, improvement in the bitter taste could be demonstrated if a microbial rennet (rennilase) was employed together with a fast acid forming starter (strain HP). Further evidence for the differential behavior of bitter and nonbitter strains may be obtained

from the following experiments. Bitter peptides prepared by the action of rennet on casein were isolated, and a medium containing these peptides as sole nitrogen source was prepared. On such a medium only nonbitter strains developed, showing that bitter strains are unable to degrade bitter peptides (Stadhouders, 1974).

However it is very likely that the importance of the specificity of the starter cultures has been greatly exaggerated. Lowrie and Lawrence (1972) have shown in a convincing manner that what really determines the development of bitterness is the rate of the proliferation of bacteria during cheese making prior to salting. If manufacturing conditions were altered so that "nonbitter strains" reached high numbers (5.3×10^9 vs. 0.28×10^9) because of low cooking temperature ($33.3°C$ vs. $38.8°C$), nonbitter strains would yield bitter cheeses. On the other hand, bitter strains in curds cooked at elevated temperatures ($39.4°C$) attaining lower counts would yield cheeses with normal taste. Inhibition of fast bitter strains by bacteriophage did lead to similar results. A model for the development of bitterness in cheddar cheese has been suggested (Fig. 19).

The previously mentioned observations would support the hypothesis that starter strains are responsible for the breakdown of the nonbitter fraction liberated during proteolysis into bitter peptides. It is not unlikely that starter bacteria are involved both in the production and degradation of the bitter peptides. A much faster rate in the formation of bitter peptides may be due to the cleavage of one to two peptide bonds in the nonbitter casein fraction, while the degradation of the bitter peptide may require the cleavage of twenty or more bonds (Jago, 1974). The situation may be even more complex since it has been recently shown that the action of rennet on casein may result in the formation of eight different bitter peptides (Green, 1977). The effect of starter density on bitter peptide formation has been also criticized. At least in the case of Gouda cheese where the cooking temperatures (30 to 35°C) permit starter organisms to reach high densities, bitter formation is not frequent (Lawrence et al., 1976). More recent work again emphasized the importance of starter strains in the formation of bitter Gouda cheeses under aseptic conditions (Kleter, 1977; Visser, 1977). Another issue

Flavor Microbiology

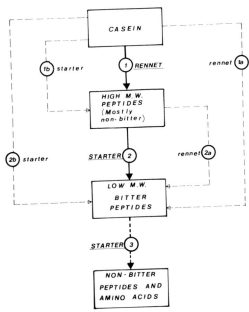

Figure 19. A Model for Bitterness Development in Cheddar Cheese. The bold arrows show the progressive sequence that is considered to be the most important. The broken arrows show steps that are thought to be less significant in contributing to bitterness intensity. After Lowrie and Lawrence (1972).

has come to the attention of workers in this area. Harwalkar and Seitz (1971) have shown that bitterness is accompanied by the phenomenon of astringency. Although the latter principle could be separated from the fraction responsible for bitterness in cheddar cheese, neither its chemistry nor its mode of formation has been sufficiently studied (Harwalkar and Elliot, 1971).

Occasionally, bitterness may also arise in yoghurt. According to Renz and Puhan (1975), this may occur if yoghurt is manufactured at lower temperatures (38°C), showing lower acidities at the beginning of storage (pH 4.4). In this case, the bitterness seems to be correlated to an increase in D-lactic acid, which may indicate that the causative organism for this defect is *Lb. bulgaricus* (and not *Str. thermophilus*) cultured at suboptimal conditions.

Bitter flavors of milk and milk products may also occur as a result of certain heat stable proteases produced by many psychotrophic organisms, mainly *Pseudomonas* sp. Although not much is known of the nature of these proteases, it is clear that the storage of products containing these enzymes may lead to undesirable off-flavors. Attempts have been made to predict the appearance of such off-flavors by measuring the tyrosine content (Hull test) of the incoming milk (White et al., 1978).

The Role of Nonstarter Microflora on Flavor Development in Cheddar Cheese

In order to gain more insight into the contribution of the so-called "adventitious bacteria" to flavor formation, it was necessary to study the problem of flavor buildup with a different approach. This was made possible by the revolutionary technique devised by Mabbit et al. (1955). Employing a chemical acidulant (δ-gluconolactone) and pasteurized milk with low bacterial counts, it is possible to go through all the stages of cheddar cheese making in a practically aseptic way. It has become quite clear that aseptic cheeses made without starter are practically devoid of cheddar flavor, while cheeses made with suitable starter organisms in the aseptic vat are similar to their counterparts made in open vats, even though in most cases the flavor of the mature aseptic cheese, with starter, was slightly milder than those of the open vat controls (Table XXII).

There have been some doubts about the suitability of delta gluconolactone as acidulant for the nonstarter (GAL) aseptic cheeses because of the fast rate of pH reduction in comparison to the lactic fermentation by starter organisms. A modified procedure has been suggested to overcome this difficulty (O'Keeffe et al., 1975).

Early publications have indicated the importance of nonstarter bacteria to cheddar flavor. Research on British and New Zealand cheddar has repeatedly emphasized the occurrence and importance of various adventitious microorganisms, e.g. nonstarter *Lb. casei* (see text earlier in this chapter), *Pediococcus* spp. (Keenan et al., 1968), and a lipolytic micrococcus (Robertson and Perry, 1961). The addition of the lipolytic micrococcus was reported to have a positive effect on cheddar flavor. The interesting fact in this work was that although these microorganisms disappear rapidly during the early period of

TABLE XXII

A COMPARISON OF THE FLAVORS OF ASEPTIC CHEESE MADE WITH
STARTER ONLY TO CONTROLS IN THE OPEN VAT

Age, Months	Flavor	
	Aseptic Cheese	*Control Cheese (nonaseptic)*
5	Cheddar flavor, some fruity off-flavor	Mild cheddar, some fruitiness
5	Very mild cheddar, bitter	Mild cheddar, slightly bitter
5	Mild cheddar, some off-flavor	Mild cheddar, a little more flavor than aseptic cheese
6	Mild cheddar, fruity off-flavor	Mild cheddar, slight off-flavor
6	Mild cheddar, sharp	Mild cheddar, sharp
6	Mild cheddar, typical sour flavor	Mild cheddar, typical sour flavor
6	Mild cheddar, slightly sharp	Mild cheddar, slightly sharp
6	Fairly strong cheddar, not full flavor	Strong, sharp cheddar
9	Mild, mature cheddar, lacking in fulness	Strong, mature cheddar
9	Mature cheddar, lacking fulness, some bitterness	Strong, mature cheddar, some bitterness
9	Mature cheddar, unpleasant fruity flavor	Unpleasant fruity-flavored cheese

Source: Adapted from Reiter et al. (1967).

ripening, their effect was noted even in the absence of viable cells. These results could be confirmed employing the aseptic vat technique with almost bacteria free milk (less than 100/ml, drawn aseptically from cows). Aseptic cheeses made with starter only had no definite cheddar flavor at maturity. The addition of culture L_1, the lipolytic micrococcus, resulted in a slight, very mild flavor. On the other hand, the presence of these micrococci could not improve the flavor of starterless GAL cheeses which generally had an unpleasant, sharp, or bitter taste without any cheddar flavor. Since the body and texture of starter only cheeses compared well with control examples, no important role was assigned to the adventitious microflora in protein breakdown (Perry and McGillivray, 1964).

The possible contribution of micrococci to cheddar flavor has been utilized for the fermentative production of a cheddar type flavor employing *Mc. caseolyticus* (Luksas, 1970b).

The "slight improvement situation" (open vats versus aseptic starter cheeses) and the question of the importance of the adventitious flora led Reiter and coworkers (1967, 1971) to introduce the "Reference Flora" technique in which microorganisms were isolated from normal curds and cheeses, propagated, and added in suitable numbers to the aseptic starter vats. Comparing these two classes of cheeses it was concluded that good cheddar can be produced in the absence of adventitious bacteria even though it may have a somewhat milder flavor. Lawrence et al., in their review on cheese starters (1976), went even further by stating that "the addition of nonstarter bacteria to cheddar cheese milk almost invariably resulted in pronounced flavor defects and at best made no difference to the basic cheddar flavor." Nevertheless, aseptic starter cheeses seem to require a somewhat longer period for maturation — up to twelve months (Reiter and Sharpe, 1971). The positive effect of micrococci on the rate of maturation may be related to their strong caseolytic activity (Nath and Ledford, 1972; Desmazeaud et al., 1974).

The effect of gram-negative bacteria in raw milk on the formation of cheese flavor has so far been studied only to a limited extent. Since lipases from such bacteria *(Achromobacter, Flavobacterium, Pseudomonas, Alcaligenes)* seem to be comparatively resistant to pasteurization temperatures, it would not be surprising that such lipases, even in the absence of viable bacteria, would contribute to the flavor buildup of cheeses. The level of these bacteria in raw milk may be critical in the formation and deterioration (rancidity) of cheese flavor (Sharpe, 1972).

Minor Components of Cheddar Flavor

Improved analytical techniques have recently revealed a number of new and interesting chemical compounds that may be related to cheese flavor. O'Keefe et al. (1969) isolated a coconut aroma fraction from the high boiling point neutral volatiles of cheddar. They identified a series of γ and δ lactones from C_{10} to C_{18} in a molecular distillate of cheese fat (Table XXIII).

The cheddar aroma of simulated cheese was enhanced by dodecalactone and tetradecalactone at 4.0 and 1.5 ppm respectively. Interestingly, the addition of lactones (C_8 to C_{14}) to margarine to improve the butter flavor was patented in 1959 (Wode

Flavor Microbiology

TABLE XXIII
SENSORY PROPERTIES OF SOME LACTONES

Compound	Flavor
γ-Decalactone	$CH_2-(CH_2)_5-CH-(CH_2)_2-C = 0$ Pleasant, fruity, peachlike odor (ring through O)
δ-Decalactone	$CH_2-(CH_2)_4-CH-(CH_2)_3-C = 0$ Oily peach odor and taste (ring through O)
γ-Dodecalactone	$CH_2-(CH_2)_7-CH-(CH_2)_2-C = 0$ Fatty, peachy odor; buttery, peachlike flavor (ring through O)
δ-Dodecalactone	$CH_2-(CH_2)_6-CH-(CH_2)_3-C = 0$ Powerful, fresh fruit, oily odor; low levels of peach pear, and plumlike flavor (ring through O)

SOURCE: After Maga (1976).

and Holm, 1959). The origin of these lactones is still uncertain. McGugan (1975) claims that lactobacilli cheese developed a more pronounced cheddar flavor but that these bacteria had little influence on the production of neutral high boiling point compounds. Milk fat containing hydroxyfatty acids may serve as precursors, hydrolysis and lactonization taking place under suitable pH and temperature conditions (Kontson et al., 1970). The possible involvement of microbes in this step has not yet been investigated, although lactone production by microorganisms

TABLE XXIV
FORMATION OF δ-DECALACTONE FROM 500 MG δ-KETOCAPROIC
ACID BY MICROORGANISMS

Organism	δ-Lactone Formed mg	$[\alpha_D]$
Cladosporium butyri	71	−15.0°
Saccharomyces fragilis	229	+48.5°
Candida globiformis	233	+55.8°
Candida pseudotropicalis	252	+48.2°
Sarcina lutea	302	−29.2°
Saccharomyces cerevisiae	355	+48.2°

SOURCE: After Muys et al. (1962).
NOTE: Note different optical activity of product.

has been definitely demonstrated in other systems (Table XXIV).

Another group of compounds recently detected in cheddar and other cheeses may also contribute to cheese flavor. Very low concentrations of pyrazines were found in distillates of old cheddar. These compounds have a nutty to peppery flavor and may collectively affect the sharpness of cheeses considering their very low odor thresholds (Sloot and Harkes, 1975; Lin, 1976).

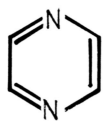

THE PYRAZINE NUCLEUS

The following pyrazine derivatives have been so far described:

2-acetyl pyrazine
2-methoxy-3-ethyl pyrazine
2,5 (or 2,6)diethyl-3-methyl pyrazine
Dimethyl pyrazine
3-ethyl, 2,6 (or 2,5)dimethyl pyrazine
Trimethyl pyrazine, tetramethyl pyrazine
Ethyl, trimethyl pyrazine.

Still, little is known about the biogenesis of these compounds. It is possible that they are reaction products of sugar metabolites and amino acids during ripening, with direct or indirect participation of microorganisms. Several bacteria have been shown to be involved in the formation of certain pyrazines. Demain et al. (1967) described a mutant of *Corynebacterium glutamicum* which accumulated high amounts of tetramethyl pyrazine in the presence of suitable amino acid precursors. Kosuge et al. (1971) studied the biosynthesis of tetramethyl pyrazine in fermented soybean products. They suggested that these compounds were formed by certain bacteria (*B. subtilis* and *B. natto*) from acetoin:

$$+\text{NH}_3$$

$$\text{CH}_3\text{-CH(OH)-CH}_3 \xrightarrow[\text{-H}_2\text{O}]{} \left[\underset{\underset{\text{NH}_2}{|}}{\text{CH}_3\text{-CH}}\text{—CO-CH}_3 \rightleftharpoons \underset{\underset{\text{NH OH}}{|\ \ |}}{\text{CH}_3\text{-C}\text{—C-CH}}\right]$$

Their possible occurrence in milk has also been suggested (McGugan, 1975). Whether similar reactions also take place during cheese ripening remains to be elucidated.

Pyrazines are probably also involved in certain off-flavors in certain dairy products. Thus, a musty potato aroma was detected in milk and other refrigerated foods. This was traced back to the metabolic activities of *Pseudomonas taetrolens* (= *Ps. graveolens*) and the formation of 2,5-dimethylpyrazine and 2-methoxy-3-isopropyl pyrazine (Morgan, 1976). In the French cheese, Gruyère de Comté, a potatolike off-flavor was described to be due to the presence of 2-methoxyl alkylpyrazines as well as the related compound, 3-methoxyl-2-propylpyridine, although details about their origin are not yet available (Dumount et al., 1975). A potatolike off-flavor has been earlier observed in cheeses made from milk containing *Ps. mucidolens* (Pinheiro et al., 1965). The nature of this off-flavor was not disclosed but may be of similar chemical composition.

New efforts for the identification of the chemical compounds primarily responsible for cheddar aroma as obtained in suitable distillates have focused on the sulphur compounds appearing during cheese ripening. The conclusion of early reports on the importance of hydrogen sulfide and the ratio of FFA/H₂S were not generally accepted by later workers (McGugan et al., 1968). This problem was again taken up by Manning and coworkers (1976, 1977) who employed the headspace gas analysis for the determination of volatile sulfur compounds. Using the aseptic vat technique, a number of cheeses were prepared and analysed. It was again confirmed that only whole milk starter cheeses developed normal cheddar flavor, while starterless or skim milk starter cheeses had little or no flavor. Among the sulphur compounds, methanthiol (CH₃SH, methyl mercaptan) was found only in cheeses with characteristic cheddar flavor (highest concentration: 19.2 nanograms/5 ml headspace). Hydrogen sulfide and dimethylsulfide (CH₃S • SCH₃), on the other hand, could be detected in both good and poor cheeses. Methanthiol is undoubtedly derived from sulfur containing aminoacids through protein degradation. If so, the bacterial activities that lead to the breakdown of milk proteins and liberation of amino acids play an important role in the formation of cheddar aroma, contrary to earlier concepts (Day, 1967).

Production of volatile sulfur compounds by microorganisms has been reviewed by Kadota and Ischida (1972). While a limited group of microorganisms may produce hydrogen sulfide through the desulfhydration of organic S compounds or direct reduction of sulfates, a large variety of organisms, gram-positive and gram-negative, aerobic and anerobic bacteria, as well as a number of eucaryotes, may produce methanthiol and dimethyl sulfide from methionine:

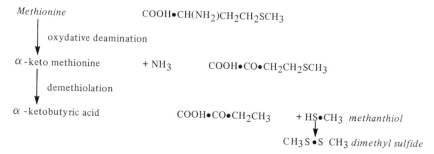

If methanthiol is essential for cheddar flavor, why do cheeses prepared from skim milk lack flavor? The answer to this question probably lies in the fact that methanthiol is lipid soluble and practically water insoluble. Milk with no fat cannot hold methanthiol, which is lost during the ripening process. The importance of milk fat for the trapping of volatile aroma compounds has been pointed out by Foda et al. (1974). Recent

research emphasizes more and more the importace of the physicochemical aspects in the ripening process of cheeses. The importance of the redox potential has recently been pointed out.
GAL cheeses prepared aseptically without starter have a redox
potential as high as +300 mV, compared to the reducing conditions prevailing in normal cheese: −150 to −200 mV (Law et al.,
1976). At high redox values, methanthiol will rapidly be oxidized
to dimethyl sulfide leading to poor flavor. These observations
further strengthen the early hypothesis regarding the importance of SH-groups for cheddar flavor (Kristoffersen, 1967). In
addition to the many metabolic activities of cheese microflora
described so far, the contribution of bacteria to the establishment of reducing condition during cheese ripening must not be
overlooked. Differences in the oxidation/reduction potential of
various streptococci and lactobacilli have been described by Carini and coworkers (1974).

FLAVOR FORMATION IN MOLD RIPENED CHEESES

Cheddar cheeses are very popular in English-speaking countries, so it is not surprising that much work in the area of cheese
flavor was done with cheddar. It is very likely that many of the
metabolic processes that contribute to cheddar flavor also occur
with other types of cheeses that undergo ripening. Nevertheless,
the fact that a number of important cheeses are mold ripened
probably overshadows much of the basic flavor-forming processes we have encountered in cheddar.

Mold ripened cheeses are either semihard (Roquefort, Gor-
gonzola, Stilton, blue) or soft types (Camembert, Brie). In the
blue type cheeses, fungal spores are usually introduced during
curd preparation, while with the soft types, surface inoculation is
also employed. The blue-green colored spores of *Penicillium
roqueforti* cause the inside of the cheeses to assume a mottled or
marbled appearance (blue-veined cheese). A high salt content (4
to 5%, i.e. approximately 10% of the aqueous phase) is also a
characteristic feature of this type. After the cheese is manufactured, long, slender needles are used to puncture the loaf in
order to admit air which encourages fungal growth. Ripening at
7 to 9°C requires several months. Mold growth becomes apparent after 8 to 10 days, reaching a maximum within 30 to 90 days.

The suitability of *P. roqueforti* to this process as proved by the very old tradition is probably due to its relative salt tolerance and good sporulation which takes place in the oxygen poor environment within the cheeses. In spite of the puncturing, oxygen depletion in the interior is quite rapid. Another reason may be due to the fact that contrary to many other molds attacking casein, *P. roqueforti* produces only very little isovaleric acid which has a disagreeable, rancidlike odor (Kowalewska et al., 1971).

Cheeses manufactured without wax cover undergo additional microbial changes due to the appearance of a slimy, reddish orange growth consisting of micrococci and rods (*Bacterium linens* or *Bacterium erythrogenes*). Some authors believe that slimy cheeses have a somewhat finer flavor than waxed cheeses on which slime formation does not take place. We shall discuss the microbial activity of these organisms in a later section.

A Norwegian version of blue cheese is Gammelost made from skim milk (starter without rennet, curdling by heating, no salting). *Mucor* or *Rhizopus* molds (*M. racemosus, M. mucedo, R. nigricans*) are spread on the surface, while in the interior the *Penicillium* (*P. roqueforti* or *P. frequentans*) develops.

Cheeses with only mold surface ripening like Camembert are characterized by their small weight. They have a soft and buttery texture (soft cheese) with a slightly salty, slightly rancid-bitter taste, appreciated by the connoisseur. They have a high water content (about 50%) but a somewhat rigid crust covered by the growth of *P. camemberti* and its greyish conidia. The high water content of such cheeses also allows other organisms like yeasts, *Geotrichum candidum,* and certain bacteria to develop before *P. camemberti* becomes the predominant organism. Both ripening (weeks) and shelf life of the soft cheeses are much shorter.

No doubt many of the flavor components of mold ripened cheeses are due to the metabolic activities of these fungi. Cheeses prepared in a similar way without molds will not develop the typical flavor of mold cheeses. Not all types of cheese have been studied sufficiently. However, the information available from a number of cheeses allows us to draw a general picture of the formation of flavor in mold ripened cheeses.

The proteolytic and lipolytic enzymes of the fungal organisms occupy a key position in mold ripened cheeses. The transformation of the initial elastic curd into the flavor rich cheese with a semisoft or soft texture concurs with mycelial growth. Casein breakdown initiated by the coagulating agent and starter enzymes is considerably accelerated by fungal proteases. Using the aseptic vat technique, casein breakdown due to mold enzymes was definitely demonstrated (Gripon et al., 1977). However, more delicate analysis has shown that the proteases of *P. roqueforti, P. caseicolum* (a colorless variant of *P. camemberti*), although yielding high nonprotein nitrogen values, had little effect on the liberation of free amino acids. Hence, many of the amino acids detectable in ripened cheeses may be due rather to the peptidase activities of starter and nonstarter organisms.

An excellent review on the enzymic activities of *P. roqueforti* has recently been published and should be consulted by the interested reader (Kinsella and Hwang, 1976). A proper balance between proteolysis and lipolysis seems to be of utmost importance in blue cheese ripening. Excessive proteolysis may be detrimental. Usually about 20 to 40 percent of the casein are converted to soluble nonprotein N compounds. If over 50 percent is hydrolyzed, the resulting cheese is soft with a pasty flavor and an unpleasant bitter taste. This may be controlled by low temperature curing and high salt concentrations which inhibit proteolysis, as well as the use of suitable strains. There seems to be a considerable variability in the proteolytic potential of different strains. A negative correlation between proteolysis and lipolysis in strains of *P. roqueforti* has been pointed out (Sato et al., 1966; Niki et al., 1966). On the other hand, too little proteolysis may result in a tough, dry, and crumbly cheese. Strains with high lipolytic activity seem to produce better cheeses in a shorter time. Hence lipolysis seems to be the more important factor in the production of such mold ripened cheeses (Niki et al., 1966).

There is general agreement about the contribution to flavor formation by the lipolytic activity of mold growth in the interior or on the surface of cheeses. Ripe blue cheeses have a very high FFA content. Values between 23.500 and 66.700 mg/kg cheese as compared to less than 2000 mg/kg in cheddar have been reported (Day, 1966). These high FFA figures for mold ripened cheeses permit us to neglect the smaller lipolytic activities of other systems (milk lipases, starter lipases). Mold lipases have been studied extensively. Contrary to milk lipases and other animal lipases, fungal lipases (*Aspergillus* sp.) seem to be more specific in the splitting of lower chain fatty acids from the milk fat. Of the milk lipases, 60 percent of the FFAs were lauric, C_{12}, and higher, while fungal lipases yielded over 99 percent of C_{10} and lower FAs (Khan et al., 1967). However, the analysis of FFA of cheeses as compiled by Day (1966) do not, in general, show any marked specificity with regard to the types of the FFA formed during the ripening; the ratios prevailing in milk fat usually also exist in blue cheeses.

The importance of FFA in cheese flavor is twofold. First, FAs per se are flavorful and secondly, they constitute precursors for the methyl ketones which contribute to the pleasant moldy flavor of the cheese. Although these have a somewhat limited importance in cheddar (see previous text), all workers seem to agree on their contribution to flavor in mold ripened cheeses. Identification and quantitation of ketones in blue cheeses were carried out by Anderson and Day (1966). High concentrations of 2-pentanone (3.6 to 20.9 mg/kg), 2-heptanone (17.6 to 71.8 mg/kg) and 2-nonanone (13.9 to 88.3 mg/kg) were found. All these values are considerably higher than the flavor threshold values of these ketones (see previous text). No direct relationship with the amount of precursors present in milk fat could be established (high 2-heptanone in mature cheese versus low octanoic acid in milk fat). This selectivity may be attributed to the curing activity of the mold. The C_4, C_6, C_{10}, and C_{12} are converted to the corresponding methyl ketones much less readily. Methyl ketones and particularly 2-heptanone have an aroma typical of mold ripened cheeses.

The biogenesis of methyl ketones undergoes the following steps (Hawke, 1966):

1. Liberation of FFA from milk fat.
2. Oxidation of FFA to form the corresponding β-keto acids (via Keto acyl CoA).
3. Decarboxylation to methyl ketones (2-alkanones).
4. Reduction to secondary alcohols.

Gehring and Knight (1963) claimed that the methyl ketone synthesizing activity is confined to the conidia and is not a function of hyphal cells. Kubeczka (1968) believes that the accumulation of methyl ketones is due to the fact that spores are unable to further metabolize the ketones via β-oxidation, which hyphal cells do. A differential biochemical activity between germinating spores and mycelial hyphae has been also noted by Lewis (1971). Ketone production by germinating spores of *A. niger* was stimulated by glucose while the activity by hyphae was suppressed. However, there are good reasons to believe that both structures partake in the formation of methyl ketones.

During the β-oxidation, actually a β-oxoacyl-CoA derivative is formed, which is deacylated and subsequently decarboxylated to the corresponding methyl ketone (Lawrence and Hawke, 1968).

Conversion of methyl ketones into secondary alcohols by growing microbial cultures has been demonstrated (Table XXV).

Since these alcohols have flavor characteristics similar to those of the corresponding methyl ketones and since the concentrations at which they occur in mature cheese are small (< 10 ppm), their contribution to mold ripened cheese flavor is considered secondary.

Free fatty acids have a significant inhibitory effect on the enzymes involved in the formation of the methyl ketones. Since the spore envelope consisting of four distinctive layers may be refractory to the very high concentrations of FAs in the cheese (Fan et al., 1976), it is conceivable that under these conditions the spore may play a greater role in the formation of these ketones.

TABLE XXV

REDUCTION OF SOME METHYL KETONES BY MICROORGANISMS

Reaction	Mycoderma	Torula sphaerica	Geotrichum candidum	Penicillium roqueforti	B. linens	Str. lactis
2-pentanone to 2-pentanol	+	+	+	+	−	−
Acetone to 2-propanol	+	+	+	+	−	−

SOURCE: Adapted from Anderson and Day (1966).

At very low concentrations, fatty acids are completely oxidized to CO_2, and little or no methyl ketones are formed. A steady supply of fatty acids is therefore necessary for flavor formation. This led Jolly and Kosikowski (1975) to the addition of fungal lipases to the milk for improved flavor production during the curing process. Increased ketone production in Roquefort and similar mold ripened cheeses has been shown to take place when yeasts (*Torulopsis severinii* or *T. minor*) were introduced into the milk. Such cheeses received higher flavor scores (Kriventsova, 1971).

Volatile and nonvolatile amines have also been detected in mold ripened cheese. Ney and Wirotama (1972) detected ethyl, butyl, isobutyl, isopentyl, dimethyl, diethyl, dipropyl, dibutyl, and hexyl amines at concentrations of 0.64 to 1.3 mg/10g cheese. Among the nonvolatile amines, tryptamine, histamine, and tyramine (up to 30 mg/10g cheese!) were identified. Amines are probably produced by the decarboxylation of free amino acids under conditions of low oxygen tension where oxidative deamination is retarded.

$$C_6H_5 (OH)-CH_2-CH-COOH \longrightarrow C_6H_5 (OH)-CH_2CH_2-NH_2 + CO_2$$
$$\underset{NH_2}{|}$$

TYROSINE   TYRAMINE

The microbial nature of these reactions, as well as their importance in cheese flavor, has not yet been elucidated.

The flavor profile of Camembert cheese with regard to the methyl ketone spectrum is quantitatively different from that of blue cheeses. Schwartz and Parks (1963) examined ketone distribution in a number of European and U.S. Camembert cheeses and found 2-nonanone to be the predominating methyl ketone (1.38 to 3.37 μmoles C_9/10 g cheese fat versus 0.60 to 1.53 μmole C_7/10 g fat). Moinas and his group (1973) also found higher values of 2-nonanone in Camembert compared to Roquefort cheeses. Another volatile compound, oct-1-en-3-ol, has been also found to contribute to the pleasant moldy flavor of Camembert and other mold ripened cheeses (Moinas et al., 1973). A concentration of 10 ppm seems to be most desirable. These compounds are produced by various molds also in nondairy media

(Kaminski et al., 1974). It is assumed that oct-1-en-3-ol is formed by the oxidation of linoleic acid, which occurs abundantly in the mycelium and spores of *P. roqueforti* (Fan and Kinsella, 1976). Attempts have been made to utilize the biochemical activities of these molds in order to prepare concentrates with strong mold cheese flavor to be used for salad dressing, snacks, and appetizers. A submerged culture process with *P. roqueforti* and a milk based medium with enzymatically hydrolized milk fat was employed. Within twenty-four to seventy-two hours of fermentation with low aeration, a ketone content seven to twelve times higher than in commercial blue cheese could be secured (Nelson, 1970). A similar process using microbial lipases and mold spores was also reported (Dwivedi and Kinsella, 1974; Jolly and Kosikowski, 1975).

Processes for the accelerated ripening of cheeses and other milk products by means of enzymic preparations derived from a variety of fungi have also been described and protected by patents. *Fomitopsis pinicola, Irpex lacteus,* and *Lenzites saepiaria* (Improvement in or Relating to Milk Foodstuffs, 1969) or *Trametes sanguinea* (Hori and Shimazono, 1973) were cultivated in suitable media under submerged conditions and the enzymes were recovered from the liquid or fermentation solids. When added to "green" cheese, a pronounced ripened cheese flavor could be established in a matter of hours. Unfortunately, no information is yet available on the chemical entities involved in such accelerated cheese curing.

SURFACE RIPENED CHEESES WITHOUT MOLD

Among the nonmold, surface ripened, semisoft cheeses, we shall briefly mention the development of flavor in Limburger cheese. The small rectangular curds (about 45% water) are rubbed with dry salt and incubated at the comparatively high temperature of 15 to 20°C. Within a few days, a heavy, reddish slime develops. After ten to fifteen days, the cheese is wrapped and stored at 10°C where ripening continues until consumption (six to eight weeks). The curing rooms supply the necessary microflora thus usually making inocula superfluous.

The first organisms to develop on the surface of the fresh blocks are probably aerobic, salt tolerant film yeasts and sometimes also some *G. candidum*. After a few days, slime develops containing a mixture of bacteria that rapidly give rise to the typical orange red *Brevibacterium linens* (family: Brevibacteriaceae, or closely related *Br. erythrogenes*), short, grampositive, non-spore-forming rods. While fresh curds have a pH of around 5.0, the activity of surface yeasts and other organisms present at the beginning of the ripening period leads to a rapid reduction in acidity, mainly due to consumption of lactic acid, leading to a pH increase to 6.0 or even higher. At this pH, vigorous growth of *Br. linens* takes place. This is accompanied by a change in cheese texture to a soft, smooth, and waxy product. The change is gradual, leading from the surface toward the interior of the cheese.

The strong proteolytic activity of *Br. linens* is decisive for the typical, putrid Limburger flavor, comprising ammonia, indole, and hydrogen sulfide. Lipolytic activity of *Br. linens* has not been recorded. It is possible that some lipolysis takes place due to the activity of the first group of surface microflora. Unfortunately, detailed studies on the chemical components of this cheese flavor and their mode of formation are still lacking.

Szumski and Cone (1962) have analyzed the microflora colonizing the fresh curds (in a similar Trappist cheese) and found a mixed culture of oxidative yeasts: *Debaryomyces* sp. and *Trichosporon* sp. It looks as if the *Trichosporon* organism also takes part in the proteolytic breakdown of the casein. This is probably due to an endoproteinase that becomes active during the decline of the number of viable yeasts and may be a preparatory stage for the overgrowth of the red-pigmented bacteria. In the same type of cheeses, hydrogen sulfide and methanthiol (1.9 and 2.3 ppm respectively) were found after sixty days' ripening. Free methionine was shown to be the precursor of methanthiol but not of H_2S. Surface organisms must also be concerned with this reaction (Grill et al., 1966a,b).

Lactose-fermenting yeasts have also been employed in cheese starters *(Saccharomyces fragilis, Candida pseudotropicalis)*, together with lactic organisms, in order to obtain semisoft cheeses with lower acidities, below 0.65 percent (Tsugo and Chang, 1970). *Torulopsis minor,* together with lactic starters, resulted in higher counts of lactic acid bacteria. Such cheeses received higher flavor scores, although most of the yeast died out after ten to fifteen days. The rapid establishment of a reductive atmosphere and production of some neutral volatiles may have been the cause for the improved flavor (Komova, 1970; Chang et al., 1972).

PICKLED SOFT CHEESES

Although many types of soft cheeses are known, most are consumed fresh. These have a high moisture content (55 to 80%) due to the whey retention. Among the soft cheeses that undergo ripening, we shall mention briefly the feta (Brindza) type popular in Balkan and mideastern countries (Davis, 1976). The fresh product is similar to highly salted cottage cheeses, whereas the ripened cheese is substantially different since the curing occurs in brined solutions or by dry salting. After 2 to 3 weeks of curing in ambient temperature, the cheese is stored in a cool temperature. The brined or pickled product is characterized by its smooth, creamy, soluble and sliceable body, and pleasant acetic acid, salty and mild flavor. Sheep milk is usually employed for typical feta cheeses, although blends of other types of milk are also employed. In native countries, milk is not pasteurized, and no starters are used.

The microbial activities taking place in such types of cheeses have been investigated employing controlled dairy and bacteriological conditions (Efthymio and Mattick, 1964). Using pasteurized milk, a starter culture composed of *Str. lactis* with either *Lb. casei* or *Lb. acidophilus* was necessary to develop the characteristic organoleptic properties of feta cheeses. This product made without lactobacilli lacked flavor and smoothness of body, whereas cheese made without streptococci was extremely soluble and gassy. Chemical analysis indicated that the rancid flavor was associated with the fatty acid content (C_2 to C_{10}). Changes during the curing process have not been recorded. Information about the microbial activities in native feta cheese made with raw milk are not yet available. However, in similar cheeses produced from raw milk in the Rhodope region of Bulgaria, a major role in the formation of the typical flavor was attributed to the presence of a variety of micrococci (*Micrococcus candidus*, *Mc. freudenreichii*, *Mc. luteolus*, *Mc. flavescens*, and *Mc. subflavescens*) which impart also a creamy yellowish color to the cheese (Gruev, 1969).

FLAVOR FORMATION IN SWISS TYPE CHEESES

Hard cheeses of the Swiss type (Emmental, Swiss, Gruyère) may be characterized by their elastic body, smooth texture with uniform smooth eyes, and a unique, nutlike *sweet* flavor (Langsrud and Reinbold, 1973). From the microbiological point of view, it is the development of the propionic acid fermentation that distinguishes Swiss types from all other cheese varieties.

The propionic acid fermentation is mediated by the propionic acid bacteria (Propionibacteriaceae: gram-positive, non-spore-forming rods; pleomorphic; aerotolerant anerobes that are catalase-positive!). A comprehensive review on the nutritional requirements and biochemistry of these organisms is now available (Hettinga and Reinbold, 1972). These bacteria are very common in the environment of dairy farms and in milk and do survive mild pasteurization treatments. Sherman (1921) was probably the first bacteriologist to provide evidence that the propionic acid bacteria are essential for characteristic flavor development in Swiss cheeses. They have a pH optimum for growth close to neutrality and are relatively salt sensitive. At lower pH levels (5.2), 3 percent salt retards their growth and fermentation. Their growth in milk is rather slow but may be enhanced by the incorporation of peptone or amino acids. They show very little proteolytic activity. Although propionic acid may be produced from a variety of carbon sources (including lactose and amino acids), the major source for propionic acid is lactic acid produced during the early stage of cheese manufacture. The following stoichiometry may be employed.

$$3 \text{ Lactic acid} \rightarrow 2 \text{ Propionic acid} + \text{Acetic acid} + CO_2 + H_2O$$

The 2:1 ratio of propionate/acetate may be found in broth cultures, although the ratio may vary with different carbon sources. The two lactic acid isomers are apparently not fermented at the same rate, the dextrorotatory (L+) form being preferred (Hunter and Frazier, 1961). Although pyruvate is easily converted to propionate, this is not a direct reduction but involves a complex enzymic system with biotin and B_{12} as cofactors (Hettinga and Reinhold, 1972).

The technology of Swiss cheese manufacture has evolved in such a manner as to provide the conditions necessary for an active propionic acid fermentation:

1. Large wheels or blocks provide a suitable environment for anaerobic bacteria.
2. Low salt concentrations (1 to 1.6% in product).
3. pH of curds after pressing never below 5.0, rising somewhat during the ripening process.
4. Incubation for several weeks at high temperatures (20 to 25°C), during which propionic acid bacteria may reach 10^9/g and over before cheeses are transferred to 2 to 5°C for several months' curing.

However, these conditions are also suitable for the butyric acid fermentation by clostridia, which may cause spoilage of the cheese — the "late fermentation" (Ritter et al., 1963). The use of nisin-producing streptococci for the inhibition of such clostridia (*Cl. tyrobutyricum* or *Cl. sporogenes*) as well as the addition of the antibiotic nisin itself — once advocated for the prevention of the "late split defect" — proved inefficient since this inhibited the aroma-producing organisms also. Further, the inhibition of the clostridia proved inadequate since during the two to three months of ripening before the split effect becomes noticeable, nisin is broken down by various lactic bacteria that form nisinase. Kundrat (1970), who studied this problem for many years, suggests the use of suitable starters comprising certain selected strains of *Lb. helveticus* and other lactic organisms that proved inhibitory to the clostridia.

Eye formation is another critical parameter in Swiss cheese manufacture. Although not directly concerned with flavor, eye production is a direct result of the propionic acid fermentation; the carbon dioxide evolved (see equation) accumulates in the weak spots of the curd. However, these must be of regular size (1.25 to 2.54 cm) and at correct intervals (2.54 to 7.62 cm) in order to be acceptable.

Starter Organisms

In Switzerland where the traditional cheese making process still prevails, thermophilic lactics *(Lb. helveticus, Str. thermophilus)* are usually employed for the acidification stage in connection with the high cooking temperature (50 to 54°C). In other countries where lower cooking temperatures are used (40 to 46°C), mesophilic lactic organisms are also employed. Although many countries use pasteurized milk for Swiss type cheese manufacture, most Swiss producers prefer raw milk, with some very mild heat treatment termed thermisation (56 to 65°C for 3 seconds at most). Severe heat treatment seems to adversely affect the cheese making. One of the reasons may be related to the fact that interactions between casein and whey proteins may take place, resulting in a deterioration of cheese texture and eye formation (Olsanský and Vychytová, 1965). The use of *Str. thermophilus* in nonpasteurized milk requires some adaptation since normal thermophilus cultures do not propagate well in raw milk (Ritter and Holzer, 1970). The use of mixed cultures here also seems to be advantageous, resulting in improved eye formation and cheese flavor (Jabarit, 1970). Addition of other starter cultures *(Lb. bulgaricus, Lb. casei)* have also been suggested. Although the propionic acid bacteria are quite abundant in raw milk in most places, small inocula of pure cultures are also added with the starter organisms to ensure proper cheese ripening.

Flavor Development

Propionic acid is the main determinant of Swiss type cheese flavor. Propionic acid values of 200 to 600 mg/100 g cheese and acetic acid values of 180 to 370 mg/100 g have been reported (Schormüller and Langner, 1960; Langler and Day, 1966). Some butyric acid (usually less than 10 mg/100 g cheese) is also found. Although the high propionic acid concentration is typical of Swiss cheese, it is by no means the only flavor component. The sweet sensation of such cheese has been attributed to a high proline concentration. According to Hintz et al. (1956), a minimal concentration of 2 mg/g is necessary for typical cheese flavor (Table XXVI).

Flavor Microbiology

TABLE XXVI
AVERAGE PROLINE CONCENTRATION AND FLAVOR PROFILE
IN SWISS CHEESE

	Flavor			
	None	*Mild*	*Medium*	*Pronounced*
Proline mg/100 g	30	180	230	420

SOURCE: From Hintz et al. (1956).

Although it is known that the propionic acid bacteria do not take an active part in the breakdown of casein, a specific peptidase that liberates high amounts of proline may be involved (Schormüller and Müller, 1957; Langsrud et al., 1977). Other workers believe that the sweet sensation is enhanced by certain lactobacilli and advocate the addition of *Lb. bulgaricus* to assure good taste (Biede et al., 1976, 1977). The use of the aseptic vat technique should help to resolve the problem of the nature of the microbial activity in the formation of sweet flavor in these cheeses.

Other flavor components like diacetyl, methyl ketones, and nonvolatile fatty acids have also been identified, but these do not seem to play an important role in Swiss cheeses. Reduction of aldehydes to their corresponding alcohols (ethanol, propanol, etc.) by the lactic organisms may explain small alcohol concentrations (< 10 mg/100 g) in Swiss type cheeses (Keenan and Bills, 1968). In the French version of this cheese type, Gruyère, the amount of volatile acids seems to be a little higher; this may be explained by the surface growth of slime organisms that leads to somewhat stronger flavor (Kuzdzal-Savoie and Kuzdzal, 1966). This is promoted by lower ripening temperature (12 to 18°C) and higher humidity. Under these conditions, the propionic acid fermentation is less vigorous, resulting in fewer eyes (Davis, 1976).

Among the sulfur compounds, dimethyl sulfide seems to be an important flavor ingredient; its concentration ranges between 0.05 and 0.18 ppm (the flavor threshold in milk being close to 0.024 ppm), i.e. about four times the values obtained in cheddar cheeses. There used to be some doubt about the origin of dimethyl sulfide since it could be produced from a precursor (methyl-methionine sulfonium salts) in milk by heating. The involvement of propionic acid bacteria in dimethyl sulfide formation in a milk culture has been demonstrated (Dykstra et al., 1971). Information on the presence of methanthiol and hydrogen sulfide is not yet available.

Of the eight propionibacterium species mentioned in the last edition of *Bergey's Manual* (1974), only four were derived from dairy sources. Usually, *Pr. shermanii* (now a subspecies of *Pr. freudenreichii*) and *Pr. arabinosum* (now *Pr. acidi-propionici*) are those concerned with Swiss cheese manufacture. There seems to be some difference with regard to their effect on texture and flavor formation. This is particularly notable in their ability to develop and ferment at low curing temperatures (2.8 to 7.2°C). Cheese made with *Pr. shermanii* (the more cold tolerant species) showed eye volumes that were too large and had a greater tendency to split (the "split defect") than *Pr. arabinosum* (Park et al., 1967).

OTHER FLAVOR DEFECTS IN CHEESES

In addition to the malty, fruity, and bitter off-flavors discussed in previous sections, a few less frequent flavor defects also associated with microbial activities will be briefly mentioned.

A flavor defect in Dutch Gouda cheese, reminiscent of the odor of cat urine or the flower of *Ribes sanguinarum,* has been reported in several countries (Badings, 1967). Gas chromatographic and mass spectrometric analysis has revealed the occurrence of 2-methyl-4-oxopentane-2-thiol, thought to be responsible for this off-flavor. Hydrogen sulfide produced during ripening forms an addition product with the precursor mesityl oxide (4-methyl-3-penten-2-one) which does not appear in normal cheeses.

$$(CH_3)_2C=CH-COCH_3$$

4-METHYL-3-PENTEN-2-ONE (MESITYL OXIDE) $+ H_2S \longrightarrow$

$$CH_3-\overset{\overset{\displaystyle CH_3}{|}}{\underset{\underset{\displaystyle SH}{|}}{C}}-CH_2COCH_3$$

2-METHYL-4-OXOPENTANE-2-THIOL

Although direct proof for the involvement of microbial activities in the formation of mesityl oxide is lacking, it is suspected that mesityl oxide is somehow produced by microbial activity since clean trays and meticulous sanitary conditions do prevent Ribes scent formation.

Another defect recorded in Gouda cheese was described as "phenolic." This was found to be due to the accumulation of p-cresol (threshold value in cheese about 0.3 ppm). The formation of p-cresol was traced back to microbial activity. Certain variants of lactobacilli *(Lb. casei, Lb. plantarum)* were shown to be able to degradate amino acids in such a way as to produce the off-flavor, sometimes together with large amounts of CO_2 (late gas effect) during ripening. The microflora was derived from inadequately filtered rennet (Badings et al., 1968). It is possible, however, that phenolic compounds also play a role in cheese flavor. Phenol and p-cresol were also found in cheddar (McGugan and Howsam, 1973) and interestingly were included together with guaiacol (o-methoxyphenol) in a synthetic cheddar flavor formulation (Pintauro, 1971). Anisole, 4-methylanisole, and 2.4-dimethylanisole were described in Camembert cheeses (Dumont et al., 1976).

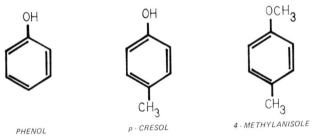

PHENOL   p - CRESOL   4 - METHYLANISOLE

MICROBIAL RENNETS AND CHEESE FLAVOR

The use of calf rennets as a milk coagulant has become problematic in recent years due to increased cheese consumption and lower availability of calf stomachs. Microbial rennets seem to be quite satisfactory with regard to their milk-clotting properties. Enzyme preparations from a variety of molds (*Endothia parasitica, Mucor miehei, Mucor pusillus* var. *Lindt*) are now available under different trade names (Huang and Dooley, 1976; Sternberg, 1976).

The question remains to what extent, if any, does the use of noncalf rennets affect the process of cheese making and how will the consumers' market react to this change in technology. It has to be pointed out that a change in technology in a traditional branch of food processing like cheese making is a very time-consuming process, often meeting with considerable reluctance by cheese makers. However, prices and difficulties in the purchase of coagulating agents exert enough pressure for the introduction of microbial rennets.

Dinesen et al. (1975) have compared the quality of cheddar cheeses at three cheese factories made with calf and microbial rennets. After fifteen months of aging, the cheddar flavor was typical in both cases. The flavor of cheeses with calf rennets scored somewhat higher during the first three months, but later (six months) the flavor differences disappeared. Microbial rennets are somewhat more proteolytic than calf rennets, leading to higher, nonprotein nitrogen in the whey. This may constitute a drawback if whey is used for nutrition. Higher thermal stability of microbial rennets may be another difficulty if whey is to be used.

The intense proteolytic activity of microbial rennets may cause problems of bitter taste, softening of texture, and loss of cheese yields (Houis et al., 1973). However, not all workers in this field noticed any bitterness in ripe cheeses (Carini et al., 1974). The bitter problem may be related to the use of certain starter organisms (described earlier) and not due to the clotting enzyme. In the United States, over 50 percent of the cheeses are manufactured with microbial enzymes. Other countries may soon follow. Attempts have also been made to use mold lipases, especially from *M. miehei*, as a substitute for the pregastric lipases of animal origin. Success has been claimed with a number of Italian cheeses (Moskowitz et al., 1977).

In this chapter, the formation of flavor, taste, and aroma in dairy products has been discussed with emphasis on the importance of microbial systems, bacteria, and fungi. Although microorganisms do play a decisive role in cheese's flavor production, it is very hard or even impossible to assess these activities without taking into consideration other enzymic and nonenzymic processes that take place in a variety of milk products and particularly those that occur during the prolonged ripening process of cheeses. Cheeses are very complex biological systems and must therefore be regarded with a holistic viewpoint. Such a system is susceptible to a variety of factors, not all of which are properly understood; but with better understanding, it is hoped that the processing of dairy products will result in superior products with desirable and constant flavor characteristics.

BIBLIOGRAPHY

Abo-Elnaga, I. G., El-Asvad, M., and Moqi, M.: Some chemical and microbiological characteristics of leben. *Milchwissensch, 32:*521, 1977.

Adda, J., Roger, S., and Dumont, J. P.: Some recent advances in the knowledge of cheese flavor. *In* Flavor of Foods and Beverages. Charalambous, G. and Inglett, G. E. (Eds.) Acad. Press, New York, 1978.

Anderson, D. E., and Day, E. A.: Quantitative evaluation and effect of certain microorganisms on flavor components of blue cheese. *J Agric Food Chem, 14:*241, 1966.

Babel, F. J.: Cottage cheese cultures. *J Dairy Sci, 42:*2009, 1959.

Badings, H. T.: Causes of Ribes flavor in cheese. *J Dairy Sci, 50:*1347, 1967.

Badings, H. T., Stadhouders, J., and Van Duin, H.: Phenolic flavor in cheese. *J Dairy Sci, 51:*31, 1968.

Barnicoat, C. R.: Diacetyl in cold-stored butters. *J Dairy Res, 8:*15, 1937.

Baroudi, A. A. G. and Collins, E. B.: Microorganisms and characteristics of laban. *J Dairy Sci, 59:*200, 1976.

Bautista, E. S., Dahiya, R. S., and Speck, M. L.: Identification of compounds causing symbiotic growth of *Streptococcus thermophilus* and *Lactobacillus bulgaricus* in milk. *J Dairy Res, 33:*299, 1966.

Biede, S. L., Reinbold, G. W., and Hammond, E. G.: Influence of *Lactobacillus bulgaricus* on microbiology and chemistry of Swiss cheese. *J Dairy Sci, 59:*854, 1976.

————: Influence of *Lactobacillus bulgaricus* on commercial Swiss cheese. *J Dairy Sci, 60:*123, 1977.

Bills, D. D. and Day, E. A.: Determination of the major free fatty acids of cheddar cheese. *J Dairy Sci, 47:*733, 1964.

Bills, D. D., Morgan, M. E., Libbey, L. M., and Day, E. A.: Identification of compounds responsible for fruity flavor defect of experimental cheddar cheeses. *J Dairy Sci, 48:*1168, 1965.

Blais, J. A., Boulet, M., Giroux, R. N., and Martin, C.: Detection of milks capable of producing fruity flavor in cheese. *Can Inst Food Sci Technol J, 11:*31, 1978.

Bottazi, V. and Vescovo, M.: Carbonyl compounds produced by yoghurt bacteria. *Neth Milk Dairy J, 23:*71, 1969.

Branen, A. L. and Keenan, T. W.: Identification of a stimulant for *Lactobacillus casei* produced by *Streptococcus lactis. Appl Microbiol, 20:*757, 1970.

————: Diacetyl and acetoin production by *Lactobacillus casei. Appl Microbiol,* 22:517, 1971.

Breene, W. M., Price, W. V., and Ernstrom, C. A.: Changes in composition of cheddar curd during manufacture as a guide to cheese making by direct acidification. *J Dairy Sci, 47:*840, 1964.

Brown, C. D. and Townsley, P. M.: Fermentation of milk by *Lactobacillus bifidus. Inst Can Food Technol, 3:*121, 1970.

Bruhn, J. C. and Collins, E. B.: Reduced nicotinamide adenine dinucleotide oxidase of *Streptococcus diacetilactis. J Dairy Sci, 53:*857, 1970.

Buchanan, R. E. and Gibbons, N. E. (Eds.): *Bergey's Manual of Determinative Bacteriology,* 8th ed. Baltimore, Williams & Wilkins, 1974.

Buyze, G., van den Hamer, J. A., and de Haan, P. G.: Correlation between hexose-mono phosphate shunt, glycolytic system and fermentation-type in lactobacilli. *Antonie Van Leeuwenhoek J Microbiol Serol, 23:*345, 1957.

Carini, S., Lodi, R., and Vezzoni, A.: I batteri lattici ed il loro potere riducente, acidificante e caseinolitico. *Latte, 2:*969, 1974.

Carini, S., Todesco, R., and Delforno, G.: Use of microbial rennets in the manufacture of soft cheese. *Latte, 2:*780, 788, 1974.

Chandan, R. C., Searles, M. A., and Finch, J.: Lipase activity of lactic cultures. *J Dairy Sci, 52:*894, 1969.

Chang, J. E., Yoshino, U., and Tsugo, T.: Studies on cheese ripened mainly with yeasts. *Jpn J Zootechn Sci, 43:*561, 1972.

Cheese and Cheese Products, Definitions and Standards. Federal Food, Drug, and Cosmetic Act 19. U.S. Food & Drug Administration, 1971.

Choisy, C.: The technological suitabilities of the thermophilic streptococci. *19th Int Dairy Congr, 1E:*422, 1974.

Chou, T. C.: "The chemical nature of the characteristic flavor of cultured buttermilk" (Ph.D. thesis, Ohio State University, 1962).

Christensen, M. D. and Pederson, C. S.: Factors affecting diacetyl production by lactic acid bacteria. *Appl Microbiol, 6:*319, 1958.

Citti, J. E., Sandine, W. E., and Elliker, P. R.: Comparison of slow and fast acid-producing *Streptococcus lactis. J Dairy Sci, 48:*14, 1965.

Clement, G., Clement, S., Begard, J., Costango, G., and Paris, R.: Hydrolysis of butter triglycerides by pancreatic lipase. *Arch Sci Physiol, 16:*237, 1962.

Cogan, T. M.: Citrate utilization in milk by *Leuconostoc cremoris* and *Streptococcus diacetilactis. J Dairy Res, 42:*139, 1975.

Collins, E.B.: Biosynthesis of flavor compounds by microorganisms. *J Dairy Sci, 55:*1022, 1972.

————: Influence of medium and temperature on end products and growth. *J Dairy Sci, 60:*799, 1977.

Collins, E. B. and Harvey, R. J.: Failure in the production of citrate permease by *Streptococcus diacetilactis. J Dairy Sci, 45:*32, 1962.

Czulak, J.: Bitter flavor in cheese. *Aust J Dairy Technol, 14:*177, 1959.

Dahiya, R. S. and Speck, M. L.: Symbiosis among lactic streptococci. *J Dairy Sci, 45:*607, 1962.

————: Identification of stimulatory factor involved in symbiotic growth of *Streptococcus lactis* and *Streptococcus cremoris. J Bacteriol, 85:*585, 1963.

Davis, J. G.: The Lactobacilli. II. Applied Aspects. *Prog Ind Microbiol, 4:*95, 1963.

————: *Cheese,* vol. III, *Manufacturing Methods.* Edinburg, Scotland, Churchill Livingstone, 1976.

Day, E. A.: Role of milk lipids in flavors of dairy products. In Gould, R. G. (Ed.): *Flavor Chemistry.* Washington, Am Chemical, 1966.

————: Cheese flavor. In Schultz, H. W. (Ed.): *The Chemistry and Physiology of Flavors.* Westport, Connecticut, Avi, 1967.

Day, E. A., Basette, R., and Keeney, M.: Identification of volatile carbonyls from cheddar cheese. *J Dairy Sci, 41:*718, 1960.

Day, E. A. and Libbey, L. M.: Cheddar cheese flavor: Gas chromatographic and mass spectral analysis of the neutral components of the aroma fraction. *J Food Sci, 29:*583, 1964.

Demain, A. L., Jackson, M., and Trenner, N. R.: Thiamine-dependent accumulation of tetramethyl pyrazine accompanying a mutation in the isoleucine-valine pathway. *J Bacteriol, 94:*327, 1967.

DeMan, J. C. and Pette, J. W.: The mechanism of diacetyl formation in butter and starters. *14th Int Dairy Congr, 7:*89, 1956.

Desmazeaud, M. J., Devoyod, J. J., Feuillat, M., and Grude, P.: Mechanism of the stimulatory action of proteolytic micrococci on lactic acid bacteria. *19th Int Dairy Congr, 1E:*424, 1974.

Dinesen, N., Emmons, D. B., Becket, D., Reiser, B., Lammond, E., and Irvine, D. M.: Cheddar cheese trials with *Mucor miehei* enzyme. *J Dairy Sci, 58:*795, 1975.

Dolezalek, J.: Proteolytic activity of *S. lactis, S. cremoris, S. diacetilactis, L. helveticus* and *L casei. Sb Vys Sk Chem Technol, 15:*59, 1967.

Drinan, D. F., Tobin, S., and Cogan, T. M.: Citric acid metabolism in hetero- and homofermentative lactic acid bacteria. *Appl Environ Microbiol, 31:*481, 1976.

Dumont, J. P., Roger, S., and Adda, J.: Mise en évidence d'un composé à hétérocycle azoté responsable d'un défaut d'arôme dans le Gruyère de Comté. *Lait, 55:*479, 1975.

————: L'arôme du Camembert: autres composés mineurs mis en évidence. *Lait, 56:*559, 1976.

Dutta, S. M., Kuila, R. K., and Ranganathan, B.: Effect of different heat treatments of milk on acid and flavor production by five single strain cultures. *Milchwissensch, 28:*231, 1973.

Dwivedi, B. K. and Kinsella, J. E.: Carbonyl production from lipolyzed milk fat by the continuous mycelial culture of *Penicillium roqueforti. J Food Sci, 39:*83, 620, 1974.

Dykstra, G. J., Drerup, D. L., Branen, A. L., and Keenan, T. W.: Formation of dimethyl sulfide by *Propionibacterium shermanii* ATCC 9617. *J Dairy Sci, 54:*168, 1971.

Efthymio, C. C. and Mattick, J. F.: Development of domestic feta cheese. *J Dairy Sci, 47:*593, 1964.

Elliker, P. R., Anderson, A. W., and Hanneson, G.: An agar culture medium for lactic acid streptococci and lactobacilli. *J Dairy Sci, 39:*1611, 1956.

Emmons, D. B., McGugan, W. A., Elliot, J. A., and Morse, P. M.: Affect of strain of starter culture and of manufacturing procedure on bitterness and protein breakdown in cheddar cheese. *J Dairy Sci, 45:*332, 1962.

Fan, T. Y., Hwang, D. H., and Kinsella, J. E.: Methyl ketone formation during germination of *Penicillin roqueforti*. *J Agric Food Chem, 24:*443, 1976.

Fan, T. Y. and Kinsella, J. E.: Changes in biochemical components during the germination of spores of *Penicillium roqueforti*. *J Sci Food Agric, 27:*745, 1976.

Foda, E. A., Hammond, E. G., Reinbold, G. W., and Hotchkiss, D. K.: Role of fat in flavor of cheddar cheese. *J Dairy Sci, 57:*1137, 1974.

Formisano, M., Coppola, S., Peruoco, G., Peruoco, S., Zoina, A., Germano, S., and Capriglione, I.: Results of five years of microbiological research on yoghurt. *Ann Microbiol, 24:*281, 1974.

Forss, D. A. and Patton, S.: Flavor of cheddar cheese. *J Dairy Sci, 49:*89, 1966.

Foster, E. M., Nelson, F. E., Speck, M. L., Doetsch, R. N., and Olson, J. C.: *Dairy Microbiology*. Englewood Cliffs, New Jersey, P-H, 1957.

Franklin, J. G. and Sharpe, M. E.: The incidence of bacteria in cheese and their association with flavour. *J Dairy Res, 30:*87, 1963.

Galesloot, T. E. and Hassing, F.: Some differences in behaviour between starters containing as aroma bacterium either *Streptococcus diacetilactis* or *Betacoccus cremoris*. *Neth Milk Dairy J, 15:*225, 1961.

Gehring, R. F. and Knight, S. G.: Fatty acid oxidation by spores of *Penicillium roqueforti*. *Appl Microbiol, 11:*166, 1963.

Gilliland, S. E., Anna, E. D., and Speck, M. L.: Concentrated cultures of *Leuconostoc citrovorum*. *Appl Microbiol, 19:*890, 1970.

Gilliland, S. E., Cobb, W. Y, Speck, M. L., and Anna, E. D.: Comparison of volatile components produced by concentrated and conventional cultures of *Leuconostoc citrovorum*. *Cult Dairy Prod J, 6:*12, 1971.

Gordon, D. F. and Speck, M. L.: Bitter peptide isolated from milk cultures of *Streptococcus cremoris*. *Appl Microbiol, 13:*537, 1965.

Green, M. L.: Milk coagulants — a review. *J Dairy Res, 44:*159, 1977.

Grill, H., Patton, S., and Cone, J. F.: Aroma significance of sulfur compounds in surface-ripened cheese. *J Dairy Sci, 49:*409, 1966a.

————: Methyl mercaptan and hydrogen sulfide as important flavor components of Trappist cheese. *J Dairy Sci, 49:*710, 1966b.

Gripon, J. C., Desmazeaud, M. J., Bars, D. L., and Bergere, J. L.: Role of proteolytic enzymes of *Streptococcus lactis, Penicillium roqueforti* and *Penicillium caseicolum* during cheese ripening. *J Dairy Sci, 60:*1532, 1977.

Grudzinskaya, E. E.: Study of Kumiss microflora. *Tr Promyshl, 26:*82, 1968.

Gruev, P.: Microflora of Rhodope Bryndza cheese. *Nauchni Tr Vissh Med Inst Sofiia , 16:*51, 1969.

Hamdan, I. Y., Kunsman, J. E., and Deane, D. D.: Acetaldehyde production by combined yoghurt cultures. *J Dairy Sci, 54:*1080, 1971.

Hammer, B. W. and Babel, F. J.: Bacteriology of butter cultures: A review. *J Dairy Sci, 26:*84, 1943.

Harper, W. J.: Chemistry of cheese flavors. *J Dairy Sci, 42:*207, 1959.

Harvey, R. J. and Collins, E. B.: Citrate transport system of *Streptococcus diacetilactis. J Bacteriol, 83:*1005, 1962.

Harwalkar, V. R. and Elliot, J. A.: Isolation of bitter and astringent fractions from cheddar cheese. *J Dairy Sci, 54:*8, 1971.

Harwalkar, V. R. and Seitz, E. W.: Production of bitter flavor components by lactic cultures. *J Dairy Sci, 54:*12, 1971.

Hawke, J. C.: The formation and metabolism of methyl ketones and related compounds — a review. *J Dairy Res, 33:*225, 1966.

Hettinga, D. H. and Reinbold, G. W.: The propionic-acid bacteria — a review. *J Milk Food Technol, 35:*295, 1972.

Hintz, P. C., Slatter, W. L., and Harper, W. J.: A survey of various free amino and fatty acids in domestic Swiss cheese. *J Dairy Sci, 39:*235, 1956.

Hori, S. and Shimazono, H.: Verfahren zur beschleunigten Erzeugung eines käseartigen Aromas in gesauerter Milch bzw. deren Folgeprodukten. Deutsch. Patentant 1, 492 822, 1973.

Hosono, A., Elliot, J. A., and McGugan, W. A.: Production of ethyl esters by some lactic acid and psychrotrophic bacteria. *J Dairy Sci, 57:*535, 1974.

Houis, G., Deroanne, C., and Coppens, R.: Comparative study of the milk coagulating and proteolytic activity of animal rennet and three of its substitutes. *Lait, 53:*529, 1973.

Huang, H. T. and Dooley, J. G.: Enhancement of cheese flavor with microbial esterases. *Biotechnol Bioeng, 18:*909, 1976.

Hunter, J. E. and Frazier, W. C.: Gas production by associated Swiss cheese bacteria. *J Dairy Sci, 44:*2176, 1961.

Improvement in or relating to milk foodstuffs. Brit. Pat. 1,169,143, 1969.

Irvine, O. R., Bullock, D. H., and Sprule, W. H.: Flavor development in pasteurized milk cheddar cheese. I. The effect of inoculating milk with *Geotrichum candidum. J Dairy Sci, 37:*637, 1954.

Jabarit, A.: Influence de la congélation et de la cryodessiccation qui sensuit sur les taux de survie et la pourcentage des deux ferments lactiques (Culture mixte). *Lait, 50:*391, 1970.

Jago, G. R.: Control of the bitter flavour defect in cheese. *Aust. J Dairy Technol, 29:*94, 1974.

Johns, C. K. and Cole, S. E.: Lactobacilli in cheddar cheese. *J Dairy Res, 26:*157, 1959.

Jolly, R. and Kosikowski, F. V.: Blue cheese flavor by microbial lipases and mold spores utilizing whey powder, butter and coconut fats. *J Food Sci, 40:*285, 1975.

Jönsson, M., and Pettersson, H. E.: Studies on the citric acid fermentation in lactic starter cultures with special interest in α-acetolactic acid. 2. Development of analytical procedures and metabolic studies. *Milchwissenschaft, 32:*587, 1977.

Kadota, H. and Ischida, Y.: Production of volatile sulfur compounds by microorganisms. *Ann Rev Microbiol, 26:*127, 1972.

Kaminsky, E., Stawicki, S., and Wasowicz, E.: Volatile flavor compounds produced by molds of Aspergillus, Penicillium and *Fungi imperfecti*. *Appl Microbiol, 27:*1001, 1974.

Kandler, O.: Stoffwechsel der Säurewecker-organismen. *Milchwissensch, 16:*523, 1961.

Keen, A. R., Walker, N. J., and Peberdy, M. F.: The formation of 2-butanone and 2-butanol in cheddar cheese. *J Dairy Res, 41:*249, 1974.

Keenan, T. W. and Bills, D. D.: Volatile compounds produced by *Propionibacterium shermanii. J Dairy Sci, 51:*797, 1968.

Keenan, T. W., Lindsay, R. C., Morgan, M. E., and Day, E. A.: Acetaldehyde production by single-strain lactic streptococci. *J Dairy Sci, 49:*10, 1966.

Keenan, T. W., Parmelee, C. E., and Branen, A. L.: Metabolism of volatile compounds of *Pediococcus cerevisiae* and their occurrence in cheddar cheese. *J Dairy Sci, 51:*1737, 1968.

Keeney, M. and Day, E. A.: Probable role of the Strecker degradation of amino acids in development of cheese flavor. *J Dairy Sci, 40:*874, 1957.

Khan, I. M., Chandan, R. C., Dill, C. V., and Shahani, K. M.: Hydrolysis of milk fat by microbial lipases. *J Dairy Sci, 49:*700, 1966.

Khan, I. M., Dill, C. W., Chandan, R. C., and Shahani, K. M.: Production and properties of the extracellular lipase of *Achromobacter lipolyticum. Biochim Biophys Acta, 132:*68, 1967.

Kinsella, J. E. and Hwang, D. H.: Enzymes of *Penicillium roqueforti* involved in the biosynthesis of cheese flavor. *CRC Crit Rev Food Sci Nutr, 8:*191, 1976.

Kleter, G.: The ripening of Gouda cheeses made under strictly aseptic conditions. *Neth Milk Dairy J, 31:*137, 1977.

Komova, E. T.: Effect of yeast No. 304 Torulopsis new type and No. 642 *Torulopsis minor* on development of lactic acid bacteria. *Tr Vologodskii Molochnyi Inst, 60:*13, 1970.

Kontson, A., Tamsma, A., Kurtz, F. E., and Pallansch, J. M.: Method for separating volatiles from steam-deodorized milk fat into lactones, free fatty acids, and nonacidic compounds. *J Dairy Sci, 53:*410, 1970.

Kosikowski, F. V.: The liberation of free amino acids in raw and pasteurized milk cheddar during ripening. *J Dairy Sci, 34:*225, 1951.

———: *Cheese Flavor*. From symposium on Chemistry of Natural Food Flavors. Quaterm Res & Engineer Center, 1957, Natick, Massachusetts.

———: *Cheese and Fermented Milk Foods*, 2d ed. Ann Arbor, Michigan, Edward Brothers, Inc., 1977.

Kosikowski, F. V. and Mocquot, G.: *Advances in Cheese Technology*. FAO Agric Studies, *38,* Rome, FAO, 1958.

Kosuge, T., Zenda, H., Tsuji, K., Yamamoto, T., and Narita, H.: Studies on flavor components of foodstuffs. *J Agric Biol Chem, 35:*693, 1971.

Kothari, S. L. and Nambudripad, V. K. N.: Isolation and identification of stimulatory substances involved in the associative growth of cheese cultures. *J Dairy Sci, 56:*423, 1973.

Kowalewska, J., Poznanski, S., and Jaworski, J.: Production of free fatty acids by lactic acid bacteria and moulds on casein medium. *Lait, 51:*421, 1971.

Kristoffersen, T.: Interrelationships of flavor and chemical changes in cheese. *J Dairy Sci, 50:*279, 1967.

Kristoffersen, T. and Gould, I. A.: Changes in flavor quality and ripening products of commercial cheddar cheese during controlled curing. *J Dairy Sci, 43:*1202, 1960.

Kriventsova, V. F.: Carbonyl compounds in Roquefort cheese. *Izv Vysshikh Uchebnykh, 3:*27, 1971.

Kubeczka, K. H.: Comparative studies on the biogenesis of volatile secondary products. II. Molds. *Arch Mikrobiol, 60:*139, 1968.

Kuila, R. K. and Ranganathan, B.: Ultraviolet light induced mutants of *Streptococcus lactis* subspecies *diacetylactis* with enhanced acid or flavor producing abilities. *J Dairy Sci, 61:*379, 1978.

Kundrat, W.: Zur Bekämpfung der Spätblähung bei Hartkäsen auf biologischen Wege. *Alimenta, 10:*133, 167, 1971.

Kupsch, H. J.: Das Bioghurt-Biogardverfahren zur Herstellung von Sauermilcherzeugnisse mit optimalen Eigenschaften. *Dtsch Molkereizeit, 93:*925, 1972.

Kuwabara, S.: Fermented milk. Brit. Pat. 1,197,257, 1970.

Kuzdzal-Savoie, S. and Kuzdzal, W.: Les acides gras libres du fromage. *Lait, 47:*9, 1966.

Langeveld, L. P. H.: Development and application of a method of detecting malty-flavour producing Streptococci in cheese. *Neth Milk Dairy J, 29:*135, 1975.

Langler, J. E. and Day, E. A.: Quantitative analysis of the major free fatty acids in Swiss cheese. *J Dairy Sci, 49:*91, 1966.

Langsrud, T. and Reinbold, G. W.: Flavor development and microbiology of Swiss cheese — a review. *J Milk Food Technol, 36:*487, 1973.

Langsrud, T., Reinbold, G. W., and Hammond, E. G.: Proline production by *Propionibacterium shermanii P59. J Dairy Sci, 60:*16, 1977.

Law, B. A., Castanon, M., and Sharpe, M. E.: The effect of non-starter bacteria on the chemical composition and the flavour of cheddar cheese. *J Dairy Res, 43:*301, 1976.

Lawrence, R. C.: Hydrogen sulphide in cheddar cheese; its estimation and possible contribution to flavour. *J Dairy Res, 30:*235, 1963a.

————: Formation of methyl ketones as artefacts during steam distillation of cheddar cheese and butter oil. *J Dairy Res, 30:*161, 1963b.

Lawrence, R. C., Creamer, L. K., Gilles, J., and Mortley, F. G.: Cheddar cheese flavor. I. The role of starters and rennets. *NZJ Dairy Sci Technol, 7:*32, 1972.

Lawrence, R. C. and Hawke, J. C.: The oxidation of fatty acids by mycelium of *Penicillium roqueforti. J Gen Microbiol, 51:*289, 1968.

Lawrence, R. C. and Pearce, L. E.: Cheese starters under control. *Dairy Ind, 37:*73, 1972.

Lawrence, R. C., Thomas, T. D., and Terzaghi, B. E.: Cheese starters — a review. *J Dairy Res, 43:*141, 1976.

Lewis, H. L.: Caproic acid metabolism and the production of 2-pentanone and gluconic acid by *Aspergillus niger. J Gen Microbiol, 63:*203, 1971.

Lin, S. S.: Alkylpyrazines in processed American cheese. *J Agric Food Chem,* 24:1252, 1976.

Lindsay, R. C.: Cultured dairy products. In Schultz, H. W. (Ed.): *The Chemistry and Physiology of Flavors.* Westport, Connecticut, Avi, 1967.

Lindsay, R. C., Day, E. D., and Sandine, W.E.: Green flavor defect in lactic starter cultures. *J Dairy Sci, 48:*863, 1965.

Lowrie, R. J. and Lawrence, R. C.: A new hypothesis to account for the development of bitterness. *NZJ Dairy Sci Technol, 7:*51, 1972.

Luksas, A. J.: Cream cheese flavor. U.S. Pat. 3,535,121, 1970a.

————: Cheddar cheese flavor. Brit. Pat. 1,207,289, 1970b.

Mabbitt, L. A., Chapman, H. R., and Berridge, N. J.: Experiments in cheesemaking without starter. *J Dairy Res, 22:*365, 1955.

MacLeod, P. and Morgan, M. E.: Differences in the ability of lactic streptococci to form aldehydes from certain amino acids. *J Dairy Sci, 41:*908, 1958.

Maga, J. A.: Lactones in Foods. *CRC Crit Rev Food Sci Nutr, 8:*1, 1976.

Manning, D. J., Chapman, H. R., and Hosking, Z. D.: The production of sulphur compounds in cheddar cheese and their significance in flavor development. *J Dairy Res, 43:*313, 1976.

Manning, D. J. and Price, J. C.: Cheddar cheese aroma — the effect of selectively removing specific classes of compounds from cheese headspace. *J Dairy Res, 44:*357, 1977.

Mather, D. W. and Babel, F. J.: Studies on the flavor of creamed cottage cheese. *J Dairy Sci, 42:*809, 1959.

Mayeaux, J. F., Sandine, W. E., and Elliker, P. R.: A selective medium for detecting leuconostoc organisms in mixed strain starter cultures. *J Dairy Sci, 45:*655, 1962.

McDonough, F. E., Hargrove, R. E., and Tittsler, R. P. L.: Selective plating medium for leuconostoc in mixed lactic cultures. *J Dairy Sci, 46:*386, 1963.

McGugan, W. A.: Cheddar cheese flavor. A review of current progress. *J Agric Food Chem, 23:*1047, 1975.

McGugan, W. A., Blais, J. A., Boulet, M., Giroux, R. N., Elliot, J. A., and Emmons, D. B.: Ethanol ethyl esters, and volatile fatty acids in fruity cheddar cheese. *Can Inst Food Sci Technol J, 8:*196, 1975.

McGugan, W. A. and Howsam, S. G.: Silylation of microgram samples in a gas chromatography trap. *J Chromatogr, 82:*370, 1973.

McGugan, W. A., Howsam, S. G., Elliot, J. A., Emmons, D. B., Reiter, B., and Sharpe, M. E.: Neutral volatiles in cheddar cheese made aseptically with and without starter culture. *J Dairy Res, 35:*237, 1968.

Miller, A., Morgan, M. E., and Libbey, L. M.: *Lactobacillus maltoaromaticus,* a new species producing a malty aroma. *Int J Syst Bacteriol, 24:*346, 1974.

Miller, I. and Kandler, O.: The free amino acids in fermented milks. *17th Int. Dairy Congr, 5:*625, 1966.

Mizuno, W. G. and Jezeski, J. J.: Studies on starter metabolism. IV. Effect of various substrates on the formation of acetoin by a mixed strain starter culture. *J Dairy Sci, 42:*251, 1959.

Moinas, M., Groux, M. J., and Horman, J.: Flavor of cheese. *Lait, 53:*601, 1973.

Moon, N. J. and Reinbold, G. W.: Commensalism and competition in mixed cultures of *Lactobacillus bulgaricus* and *Streptococcus thermophilus. J Milk Food Technol, 39:337,* 1976.

Morgan, M. E.: The chemistry of some microbiologically induced flavor defects in milk and dairy foods. *Biotechnol Bioeng, 18:953,* 1976.

Morgan, M. E., Lindsay, R. C., Libbey, L. M., and Pereira, R. L.: Identity of additional aroma constituents in milk cultures of *Streptococcus lactis var. maltigenes. J Dairy Sci, 49:*15, 1966.

Moskowitz, G. J., Shen, T., West, I. R., Cassaigne, R., and Feldman, L. I.: Properties of the esterase produced by *Mucor miehi* to develop flavor in dairy products. *J Dairy Sci, 60:*1260, 1977.

Mulder, H.: Taste and flavour forming substances in cheese. *Neth Milk Dairy J, 6:*157, 1952.

Muys, G. T., van der Ven, B., and deJonge, A. P.: Synthesis of optically active γ- and δ-lactones by microbiological reduction. *Nature, 194:*995, 1962.

Nakae, T. and Elliot, J. A.: Production of volatile fatty acids by some lactic acid bacteria. *J Dairy Sci, 48:*293, 1965.

Nakanishi, T. and Arai, I.: Studies on lactose fermenting yeasts. *Jpn J Dairy Sci, 22:*221, 1973.

Nath, K. R. and Ledford, R. A.: Caseinolytic activity of micrococci isolated from cheddar cheese. *J Dairy Sci, 55:*1424, 1972.

Nelson, J. H.: Production of blue cheese flavor via submerged fermentation by *Penicillium roqueforti. J Agric Food Chem, 18:*567, 1970.

Ney, K. H. and Wirotama, I. P.: Untersuchung von Edelpilzkäse Aroma. *Z Lebensm Unters Forsch, 149:*275, 1972.

Niki, T., Yoshioka, Y., and Ahiko, K.: Proteolytic and lipolytic activities of *Penicillium roqueforti* isolated from blue cheese. *17th Int Dairy Congr, D:*531, 1966.

Nurmiko, V.: Phenylalanine as a precursor in the biosynthesis of folinic acid in lactic acid bacteria. *Suomen Kemistilehti, 288:*62, 1955.

Ohren, J. A. and Tuckey, S. L.: Relation of flavor development in cheddar cheese to chemical changes in the fat of the cheese. *J Dairy Sci, 52:*598, 1969.

O'Keefe, P. W., Libbey, L. M., and Lindsay, R. C.: Lactones in cheddar cheese. *J Dairy Sci, 52:*888, 1969.

O'Keeffe, R. B., Fox, P. F., and Daly, C.: Proteolysis in cheddar cheese: influence of the rate of acid production during manufacture. *J Dairy Res, 42:*111, 1975.

Olsansky, C. and Vychytova, H.: Effect of pasteurization on the quality of Emmental cheese. *Sb Praci Vyzk Ust Mek, 1963:*83, 1965.

Orla-Jensen, S.: *The Lactic Acid Streptococci.* Copenhagen, Munksgaard, 1919.

Oterholm, A., Ordal, Z. J., and Witter, L. D.: Glycerol esterhydrolase activity of lactic acid bacteria. *Appl Microbiol, 16:*524, 1968.

Pack, M. Y., Sandine, W. E., Elliker, P. R., Day, E. A., and Lindsay, R. C.: Owades and Jakovac method for diacetyl determination in mixed strain starters. *J Dairy Sci, 47:*981, 1964.

Pack, M. Y., Vedamuthu, E. R., Sandine, W. E., and Elliker, P. R.: Effect of temperature on growth and diacetyl production by aroma bacteria in single and mixed strain lactic cultures. *J Dairy Sci, 51:*339, 1968.

Park, H. S., Reinbold, G. W., and Hammond, E. G.: Role of propionibacteria in split defect of Swiss cheese. *J Dairy Sci, 50:*820, 1967.

Patton, S., Wong, N. P., and Forss, D. A.: Some volatile components of cheddar cheese. *J Dairy Sci, 41:*851, 1958.

Pelissier, J. P. and Ribadeau-Dumas, B.: Pourqui les fromages ont-ils parfois un gout amer? *Rev Lait Franc, 341:*1, 1976.

Perry, K. D.: A comparison of the influence of *Streptococcus lactis* and *Str. cremoris* starters on the flavour of cheddar cheese. *J Dairy Res, 28:*221, 1961.

Perry, K. D., and McGillivray, W. A.: The manufacture of 'normal' and 'starter free' cheddar cheese under controlled bacteriological conditions. *J Dairy Res, 31:*155, 1964.

Pette, J. W.: Some aspects of the butter aroma problem. *13th Int Dairy Congr, 2:*572, 1949.

Pinheiro, A. J. R., Liska, B. J., and Parmelee, C. E.: Heat stability of lipases of selected psychrophilic bacteria in milk and cheese. *J Dairy Sci, 48:*983, 1965.

Pintauro, N.: *Flavor Technology.* Park Ridge, New Jersey, Noyes, 1971.

Price, W. V. and Call, A. O.: Cheddar cheese; comparison of effects of raw and heated milk on quality and ripening. *J Milk Food Technol, 32:*304, 1969.

Rapp, M.: Über das Eiweissabbauvermögen von Milchsäurebakterien. *Milchwissensch, 24:*208, 1969.

Reddy, M. S., Bills, D. D., Lindsay, R. C., Libbey, L. M., Miller, A., and Morgan, M. E.: Ester production by *Pseudomonas fragi. J Dairy Sci, 51:*656, 1968.

Reddy, M. S., Vedamuthu, E. R., Washam, C. J., and Reinbold, G. W.: Differential agar medium for separating *Streptococcus lactis* and *Streptococcus cremoris. Appl Microbiol, 18:*255, 1969.

Reiter, B., Fryer, T. W., Pickering, A., Chapman, H. R., Lawrence, R. C., and Sharpe, M. E.: The effect of the microbial flora on the flavour and free fatty acid composition of cheddar cheese. *J Dairy Res, 34:*257, 1967.

Reiter, B. and Møller-Madsen, A.: Cheese and butter starters — a review. *J Dairy Res, 30:*419, 1963.

Reiter, B. and Oram, J. D.: Nutritional studies on cheese starters. *J Dairy Res, 29:*63, 1962.

Reiter, B. and Sharpe, M. E.: Relationship of the microflora to the flavour of cheddar cheese. *J Appl Bacteriol, 34:*63, 1971.

Renz, U. and Puhan, Z.: Beitrag zur Kenntnis von Faktoren, die Bitterkeit im Joghurt begünstigen. *Milchwissensch, 30:*265, 1975.

Ritter, P. and Holzer, H.: Resistance to raw milk of lactic acid bacteria for cheese cultures. *13th Int Dairy Congr, 1E:*366, 1970.

Ritter, W., and Nussbaumer, T.: Säurewecker. *Schweiz Milchztg, 62:*431, 1936.

Ritter, W., Sahli, K. W., Schilt, P., and Heuscher, E.: Beitrag zur Kenntnis der Lactatvergärung durch Butter-säurebacillen. *Schweiz Milchztg, 89:*723, 1963.

Robertson, P. S.: Recent developments affecting the cheddar cheesemaking process — a review. *J Dairy Res, 33:*343, 1966.

Robertson, P. S. and Perry, K. D.: Enhancement of the flavour of cheddar cheese by adding a strain of micrococcus to the milk. *J Dairy Res, 28:*245, 1961.

Robinson, R. K. and Tamime, A. W.: Yoghurt — a review of the product and its manufacture. *J Soc Dairy Technol, 28:*149, 1975.

Rushing, N. B. and Senn, V. J.: Effect of citric acid concentration on the formation of diacetyl by certain lactic acid bacteria. *Appl Microbiol, 8:*286, 1960.

Sanders, G. P.: Cheeses varieties and descriptions. Agriculture Handbook 54. Washington, U.S. Govt. Print. Office, 1953.

Sandine, W. E., Daly, C., Elliker, P. R., and Vedamuthu, E. R.: Causes and control of culture-related flavor defects in cultured dairy products. *J Dairy Sci, 55:*1031, 1972.

Sato, M., Honda, T., Yamada, Y., Tanaka, A., and Kawanami, T.: A study of free fatty acids, volatile carbonyl compounds and tyrosine in blue cheese. *17th Int Dairy Congr, D:*539, 1966.

Scarpellino, R. and Kosikowski, F. V.: Evolution of volatile compounds in ripening raw and pasteurized milk cheddar cheese observed by gas chromatography. *J Dairy Sci, 45:*343, 1962.

Schmidt, R. H.: Species differences and effect of incubation time on lactic streptococcal intracellular proteolytic enzyme activity. *J Dairy Sci, 60:*1677, 1977.

Schormüller, J.: The chemistry and biochemistry of cheese ripening. *Adv Food Res, 16:*231, 1968.

Schormüller, J. and Langner, H.: Über die organischen Säuren verschiedener Käsearten. *Z Lebensm Unters Forsch, 100:*380, 1960.

Schormüller, J. and Müller, H.: Beitrage zur Biochemie der Käsereifung. XIX. Über die Bestimmung und Kennzeichnung der Prolidase in Sauermilchkäse. *Z Lebensm Unters Forsch, 105:*39, 1957.

Schwartz, D. P. and Parks, W.: Quantitative analysis of methylketones in blue cheese fat. *J Dairy Sci, 46:*989, 1963.

Seitz, E. W., Sandine, W. E., Elliker, P. R., and Day, E. A.: Studies on diacetyl biosynthesis by *Streptococcus diacetilactis. Can J Microbiol, 9:*431, 1963a.

Seitz, E. W., Sandine, W. E., Elliker, P. R., and Day, E. A.: Distribution of diacetyl reductase among bacteria. *J Dairy Sci, 46:*186, 1963b.

Sharpe, M. E.: The selective action of thallous acetate for Lactobacilli. *J Appl Bacteriol, 18:*274, 1955.

————: The relation of the microflora to the flavor of some dairy products. 3d Nordic Aroma Symposium, Finland, 1972, p.64.

Sheldon, R. M., Lindsay, R. C., Libbey, L. M., and Morgan, M. E.: Chemical nature of malty flavor and aroma produced by *Streptococcus lactis var. maltigenes. Appl Microbiol, 22:*263, 1971.

Sherman, J. M.: The cause of eyes and characteristic flavor in Emmental or Swiss cheese. *J Bacteriol, 6:*379, 1921.

Silverman, G. J. and Kosikowski, F. V.: Amines in cheddar cheese. *J Dairy Sci,* 39:1134, 1956.

Skean, J. D. and Overcast, W. W.: Another medium for enumerating citrate-fermenting bacteria in lactic cultures. *J Dairy Sci, 45:*1530, 1962.

Sloot, D. and Harkes, P. D.: Volatile trace components in Gouda cheese. *J Agric Food Chem, 23:*356, 1975.

Speck, M. L.: Market outlook for acidophilus food products. *Cult Dairy Prod J, 10:*8, 1975.

Speckman, R. A. and Collins, E. B.: Separation of diacetyl, acetoin, and 2,3-butylene glycol by salting-out chromatography. *Anal Biochem, 22:*154, 1968a.

——: Diacetyl biosynthesis in *Streptococcus diacetilactis* and *Leuconostoc citrovorum. J Bacteriol, 95:*174, 1968b.

——: Incorporation of radioactive acetate into diacetyl by *Streptococcus diacetylactis. Appl Microbiol, 26:*744, 1973.

Stadhouders, J.: Dairy starter cultures. *Milchwissensch, 29:*329, 1974.

Stadhouders, J. and Veringa, H. A.: Fat hydrolysis by lactic acid bacteria in cheese. *Neth Milk Dairy J, 27:*77, 1973.

Sternberg, M.: Microbial rennets. *Adv Appl Microbiol, 20:*135, 1976.

Swartling, P. F.: Biochemical and serological properties of some citric acid fermenting streptococci from milk and dairy products. *J Dairy Res, 18:*256, 1951.

Szumski, S. A. and Cone, J. F.: Possible role of yeast endoproteinases in surface-ripened cheeses. *J Dairy Sci, 45:*349, 1962.

Thomas, S. B., Druce, R. G., and Jones, M.: Influence of production conditions on the bacteriological quality of refrigerated farm bulk tank milk. *J Appl Bacteriol, 34:*659, 1971.

Tsugo, T. and Chang, J. E.: Studies on cheese ripened mainly with yeasts. *Jpn J Zootechn Sci, 41:*445, 1970.

Turčić, M., Rošić, J., and Conić, V.: Influence of *Str. thermophilus* and *Lb. bulgaricus* culture on volatile acids content in the flavor components of yoghurt. *Milchwissensch, 24:*277, 1969.

Turner, N., Sandine, W. E., Elliker, P. R., and Day, E. A.: Use of tetrazolium dyes in an agar medium for detecting leuconostoc organisms in mixed strain starter cultures. *J Dairy Sci, 46:*380, 1963.

Umemoto, Y.: A method for the detection of weak lipolysis of dairy lactic acid bacteria on double layered agar plates. *J Agric Biol Chem, 33:*1651, 1969.

Van Beynum, J. and Pette, J. W.: The decomposition of citric acid by *Betacoccus cremoris. J Dairy Res, 10:*250, 1939.

Vedamuthu, E. R., Sandine, W. E., and Elliker, P. R.: Influence of milk citrate concentration on associative growth of lactic streptococci. *J Dairy Sci, 47:*110, 1964.

Ven, B., van der Begemann, P. H., and Schogt, J. C. M.: Precursors of methyl ketones in butter. *J Lipid Res, 4:*91, 1963.

Visser, F. H. W.: Contribution of enzymes from rennet, starter bacteria and milk to proteolysis and flavor development of Gouda cheeses. *Neth Milk*

*Dairy J, 31:*188, 1977.

Walker, N. J. and Keen, A. R.: Formation of methyl ketones during ripening of cheddar cheese. *J Dairy Res, 41:*73, 1974.

Walsh, B. and Cogan, T. M.: Diacetyl, acetoin, and acetaldehyde production by mixed-species lactic starter cultures. *Appl Microbiol, 26:*820, 1973.

————: Further studies on the estimation of diacetyl by the methods of Prill and Hammer, and Owades and Jakovac. *J Dairy Res, 41:*31, 1974.

White, C. H., Gillis, W. T., Simmler, D. L., Galal, M. K., Walsh, J. R., and Adams, J. T.: Evaluation of raw milk quality tests. *J Food Protection, 41:*356, 1978.

Williamson, W. T. and Speck, M. L.: Proteolysis and curd tension in milk associated with accelerated starter culture growth. *J Dairy Sci, 45:*164, 1962.

Wode, N. G. and Holm, U.: Process for improving the taste and flavor of margarine and other foods and edible substances. U.S. Pat. 2,903,364, 1959.

Wong, N. P.: Cheese chemistry. In Webb, B. H., Johnson, A. H., and Alford, J. A. (Eds.): *Fundamentals of Dairy Chemistry.* Westport, Connecticut, Avi, 1974.

Yu, J. M. and Nakanishi, T.: Studies on production of flavor constituents by various lactic acid bacteria. *Jpn J Diary Sci, 24:*27, 1975.

Zalashko, M. V., Orbatsova, N. V., and Savchenko, E. I.: Proteolytic activity of lactic acid bacteria. In Samstsevich, S. Albert (Ed.): *Fiziologia i biokhimya mikroorganizmo.* Minsk, Akademia Nauk, Bssr., 1970.

Zvyaginstev, V. I., Medvedeva, Z. P., and Burov, I. P.: Quantitative determination of bitter peptides in culture media of lactic acid bacteria. *Molochnaya Promyshlemnost, 33:*17, 1972.

Chapter 4

MEAT, MICROORGANISMS AND FLAVOR

. . . that thou mayest give them meat in due season.

—Psalms 104:27

MEAT FLAVOR

THE FLAVOR OF MEAT and meat products occupies an important position in modern meat industries. Intensive animal raising and more advanced feeding regimes have made the issue of meat flavor of considerable interest to the manufacture of meat products as well as to the meat chemist. That microbes are involved in the formation of meat flavor may be surmised from the observations of Harris (1968), who examined the quality of chicken meat that had been kept gnotobiologically. Chicken raised under conventional conditions had more characteristic meat flavor than those grown aseptically. It is very likely that the microbial population of the gut synthesizes aroma products that are absorbed and fixed in various parts of fat and muscle tissues of the animal. There is still much to be learned about the flavor components of meat, those derived from the animal itself and those formed by its microbial population. It is most likely that the complex microflora of ruminant animals should affect the flavor profile of their meats. Such studies have not yet been reported.

Aging greatly influences the flavor of cooked meat. The desired flavor is not produced in cooking unaged beef; rather, a metallic and astringent flavor results. The enzymes present in raw meat continue to act on meat components during aging and contribute to the production of flavor precursors. Proteolytic enzymes degrade certain muscle protein and may contribute to flavor, although it is doubtful whether this affects the tenderization of meats (Davey and Gilbert, 1966). Usually, aging is carried out for 1 to 4 weeks at 4°C. During this period, certain compo-

119

nents are transformed, e.g. formation of inosinic acid, which enhances the flavor of cooked meat.

The microflora of meats (after slaughter) have received considerable attention. Among the nonpathogenic organisms, the following bacteria were isolated and identified — the gram-negatives coliforms, *Alcaligenes, Flavobacterium, Achromobacter, Pseudomonas, Proteus,* as well as many lactic acid bacteria (Jensen, 1954).

Ockerman and Cahil (1977) studied the effect of these organisms on sterile bovine tissues with a view of gaining more information on the metabolic activities of such organisms and their effect on the organoleptic qualities of meats. Parameters such as growth, pH, water-binding capacity, changes in color, tenderness, juiciness, flavors and off-flavors as a function of storage time at refrigeration temperatures were studied. The pH of muscle tissues inoculated with *Ps. putrefaciens* and *B. subtilis* increased considerably (7.4 and 6.3 respectively), while that with *Leuconostoc mesenteroides* fell down to 5.3 (versus 5.5 of control) after twenty-one days of storage. Taste panel evaluation of tenderness showed much higher scores for the *Pseudomonas* treatment, while the flavor score was considerably reduced by this organism and was accompanied by strong off-flavors. Mast and Stephens (1972) studied the effect of psychrophilic bacteria (10^4/g) on the flavor development in chicken breast meat during cold storage of up to fourteen days. While meats inoculated with a *Flavobacterium* species maintained the odor of "good" chicken throughout storage, meats containing a species of *Alcaligenes* had a good aroma after three days, a very slight off-flavor after seven days, and a very strong "sweet ammonia-like odor" after fourteen days. Meats inoculated with *Ps. putrefaciens* retained a good aroma through seven days but developed a repulsive odor after fourteen days of storage. Broth extracts from sterile chicken meat were judged by a taste panel to be superior in flavor to the broth of inoculated meat samples. However, broths from meats inoculated with *Ps. putrefaciens* and stored up to seven days were preferred to control samples stored for similar periods.

The chemical nature of meat flavor is a very complex one. Here it is sufficient to mention that close to 200 volatile and nonvolatile compounds have been described in red (beef, pork, lamb) as well as chicken meats. Flavor develops largely after cooking and is connected to a great extent with the water soluble fraction of the meat. Heating the protein residue after extraction of lean meat did not contribute to flavor. Among the flavor precursors of uncooked meat, amino acids and sugars that contribute to the browning reaction (Maillard) as well as the formation of aldehydes via the Strecker degradation, upon heating constitute the greatest portion of the flavor of cooked meat. The volatile fraction of a cooked meat water extract distillate contained formaldehyde, acetaldehyde, propionaldehyde, 2-methyl propanal, 3-methyl butanal, acetone, 2-butanone, methanol, ethanol, diacetyl, formic acid, propionic acid, 2-methyl propionic acid, ammonia, hydrogen sulfide, dimethyl sulfide, ethyl mercaptan, etc. However, it looks as if the major portion of meat flavor is concentrated in a high boiling point fraction (Hornstein, 1967). A detailed description of all the flavor components of meats known to date has been compiled by Herz and Chang (1970) and Dwivedi (1974).

Fats yield a number of compounds that may contribute significantly to meat flavor. Among the carbonyl compounds, deca 2, 4 dienal seems to occupy a special position. Its high concentration (0.01 μmole per 10 g fat) is about thirty times the flavor threshold in fat and yields a deep fat fried aroma (Patton et al., 1969). This compound probably evolves from linoleic acid during the heating of fat. However, it must be borne in mind that lipids affect meat flavor not only by virtue of their chemical composition but also by acting as a reservoir for fat soluble substances.

We still understand poorly the nature of the flavor differences between the various kinds of meats; are they only of quantitative nature or also qualitative? Most of the compounds found in red meats were also detected in chicken (Pippen, 1967). Some claim that the chicken flavor depends a great deal on the concentration of carbonyl compounds (Minor et al., 1965). Most likely, the subtle balance of the various aroma components in cooked meats determines the flavor character of the different types of meats.

Little is known about the contribution, if any, of the psychrophilic microflora occurring on meat parts during storage to the eventual flavor of cooked meats. An acceleration of the aging process has been suggested by covering meats with a suspension of the phycomycete *Thamnidium* sp. and holding the meat for thirty-two hours at 72°F (Tenfold Faster Flavor Improvement of Beef, 1963). We shall limit our discussion to those meat products that undergo some kind of fermentation (ham, sausages).

MEAT CURING

The curing of meats was originally used as a means of meat preservation during times of plenty for a nutritional food supply at times of scarcity. Since the advent of refrigeration, cured meats have been prepared (usually a "mild cure" together with smoking) for the elaboration of a special type of food with considerable appeal to the consumer.

The process of meat curing may be described as the treatment of meat or meat products (sausages) with salts (NaCl, usually together with nitrates or nitrites) and their subsequent maturation. While meats may be dry salted, today the brining process is generally preferred. In the case of sausages, the salts are introduced to the meat emulsion prior to the stuffing of the casings.

From the microbiological point of view, two major factors control the process of curing. First, contrary to vegetable food fermentations, the substrate is almost entirely made of proteinaceous and fatty materials with high pH (close to neutrality), likely to undergo rapid spoiling and putrefaction. There is some similarity to cheese ripening, but the latter is usually brought to somewhat lower pH values prior to maturation. Second, the salts added in order to protect the meats from undesirable proteolytic and gas-forming organisms include nitrate or nitrite for the formation of nitrosomyoglobin or nitrosohemoglobin, resulting in the stabilization of the red color (nitrosohemochrome) through the action of salts, dessication, or heat. This pigment is much less sensitive to oxidation and imparts a stable visual appearance to the product. This is especially important in the case of sausages intended for cooking (frankfurters, etc.). From the sanitary point of view, the curing salts exert the well-known protection against botulism.

Although the normal population of meats prepared from

healthy animals under hygienic conditions is quite low, the population increases dramatically with handling, especially in comminuted meats with large surface areas where bacterial proliferation takes place. In addition, various spices, usually with heavy bacterial loads, are introduced into meat emulsions for sausage production. Thus, the proteinaceous food becomes populated with a large variety of genera and species before the elective action of salts becomes established. The proteolytic and lipolytic activities of these organisms are drastically reduced when meats are brought in contact with the salt. However, the microaerophilic environment of meat curing encourages rapid reduction of nitrates to nitrites ($NO_3 \rightarrow NO_2$) by a variety of halotolerant organisms. Frequently, ascorbic acid is added in order to assure a more reductive environment and the irreversible formation of the compound responsible for color stabilization. A slightly acidic environment favors the bactericidal action of nitrite, while pH values below 5.0 inhibit nitrate reduction. The production of nitric oxide ($NO_2 \rightarrow N_2O_3 \rightarrow NO$), which leads to the formation of the nitroso compounds, seems to be spontaneous below pH 6.0 but requires a nitrite reductase of lactic organisms above pH 6.2 (Founaud, 1976). We shall not go into further details on the mechanism of nitrate reduction (Backus and Deibel, 1972). These reactions have been investigated by a number of workers (Deibel et al., 1961; Shank et al., 1961; Backus and Deibel, 1972). Here it is worth mentioning that some workers believe that muscle with stabilized color has a higher flavor value than a piece of meat with little or no nitrosohemochrome formed during cooking (Grau, 1960). However, the chemical nature of cured flavor has yet to be resolved (Piotrowski et al., 1970).

In addition to the color stabilization process of meat curing, the salt environment encourages a lactic fermentation that utilizes meat sugars, as well as added carbohydrates, for acid production and a drop in pH usually down to or even below 5.0. It is worth mentioning that many of the meat proteins have an isoelectric point around 5.3, so pH reduction is important in the gelation of the soluble components that make sausages firm and sliceable (Liepe, 1976).

The formation of nitrosomyoglobin involves a complex series

of biochemical and biophysical reactions, many of which are still poorly understood (Koizumi et al., 1971). Recent studies have shown that the amount of nitrite necessary for the color stabilization process is actually small (around 25 ppm). This has been very important from the toxicological viewpoint since the discovery of nitrosamines in cured meats (Scalan, 1975). On the other hand, the formation of the typical cured flavor requires concentrations of nitrite at least four times higher (MacDougall et al., 1975). We still do not know what chemical entities constitute the flavor of curing. It is very likely that carbonyl compounds in the lipid fraction take the greatest share. Lillard and Ayres (1969) found a large variety of carbonyl compounds (Table XXVII).

We still know very little about the formation of these aldehydes, whether they are degradation products of fatty acids during heating or whether microbial activity of the curing process is involved in their formation.

A variety of meat-curing technologies are now in use. Artery pumping, multiple injection, hot and cold cures with various holding regimes are now employed (Kramlich et al., 1973). It is very likely that different technologies favor somewhat different microfloras which may ultimately affect the flavor of the product. Sausage curing (summer sausages, dry and semidry) may also be carried out in a number of ways. The meat emulsion with

TABLE XXVII

CARBONYL COMPOUNDS IN COUNTRY CURED HAM LIPIDS

Alkanals	alk-2-enals	alk-2, 4-dienals
methanal	hex-2-enal	hept-2, 4-dienal
ethanal	hept-2-enal	nona-2, 6-dienal
propanal	oct-2-enal	deca-2, 4-dienal
n-butanal	non-2-enal	undeca-2, 4-dienal
n-pentanal	dec-2-enal	dodeca-2, 4-dienal
n-hexanal	undec-2-enal	
n-heptanal	dodec-2-enal	
n-octanal		
n-nonanal		
n-decanal		
n-undecanal		
n-dodecanal		

Source: From Lillard and Ayres (1964).

various nonmeat ingredients is prepared with 1 to 3.5 percent salt (on meat), 0.25 to 0.5 percent nitrate (with or without nitrites), and 0.5 to 1.5 percent fermentable sugars. After the stuffing of the emulsion into the casing, the sausages are either held at temperatures of 20 to 25°C for 16 to 24 hours prior to placing in a smokehouse or are put directly into the smokehouse after first being heated without smoke for 8 to 16 hours, then heated and smoked for a variable period. Heating temperatures in the smokehouse may vary from 30 to 70°C with suitable humidities, according to the speed and extent of drying desired and conforming with local regulations for the inactivation of trichinellae. In another procedure, meats are steam treated (24 to 48 hours at 20 to 30°C) and air dried (4 to 6 weeks at 10 to 15°C).

What are the microorganisms involved in the process of meat curing and what, if any, is their contribution to the development of the flavor of such meats? Most authors describe the occurrence of two main groups of organisms: (1) the micrococci, usually found at concentrations ten times higher than other bacteria, at least during the initial phases of curing and (2) the lactic organisms: *Lactobacillus, Streptococcus,* and *Pediococcus,* many authors emphasizing the occurrence of *Lb. plantarum* and *Pd. cerevisiae.*

The bacteriology of the micrococci occurring in meats is a very complex one. The gram-positive, catalase-positive, salt and nitrite tolerant bacteria appear in clusters or packets of coccoidal cells. Current systematics tend to separate the more fermentative types *(Staphylococcus)* from the more oxidative types *(Micrococcus).* Kitchell (1962) enumerates twenty-seven different species that may be isolated from cured meat products. Although the last edition of *Bergey's Manual of Determinative Bacteriology* (1974) accepts only three separate species for each genus, there seems to be quite a variety of strains with somewhat different characters. *Microbacterium thermosphactum* is frequently mentioned as an important component of the meat population (Davidson et al., 1968). However, this organism is not a micrococcus but a rather pleomorphic microbe that does not reduce nitrates, and we do not know what role, if any, this organism plays during the curing of meat.

Micrococci take an active part in the process of meat curing, not only in the reduction of nitrates but also in other reactions. Stolić (1975) studied the microflora of Yugoslavian sausage ripening (*fruskogorska*) and found that micrococci dominate the microflora only during the initial phase (up to 4 days, about 5.5 $\times$ 10^5/g) of the curing process, thereafter gradually declining and yielding their place to the lactobacilli (Fig. 20). No doubt, during this first stage micrococci make their contribution to the proteolytic breakdown of muscle proteins. Indeed, during sausage ripening the concentration of soluble nitrogen compounds increases and may reach up to 25 percent of the total nitrogen of the meat (Mihalyi and Körmendy, 1967). At least part of the liberation of free amino acids has been shown to be due to the bacterial proteases. It is most likely that micrococci also affect the nature of the lipid fraction and thus contribute significantly to the flavor formation of cured products.

Unlike the flavor of cooked meat, which is produced during heating from precursor substances, the characteristic aroma of dry or fermented sausage is related in part to the hydrolytic and oxidative changes occurring in the lipid fraction during the curing process (Cantoni et al., 1967). Micrococci are considered to be strong lipolytic organisms, probably having little specificity with regard to the nature or the position of the fatty acids (Alford et al., 1971; Debevere et al., 1976). Micrococci may liberate a large variety of fatty acids that are liable to further transformations, leading to the formation of methyl ketones and saturated and unsaturated aldehydes, which at certain concentrations may contribute to the characteristic flavor of the meat product while at other concentrations may indicate meat spoilage. These reactions are related to the formation of hydroperoxides, from which aldehydes, alcohols, and ketones may be formed (after Pezold, 1969).

Many of these reactions are known to be enzymatically

$$\text{Unsaturated Fatty Acids} \longrightarrow \underset{\underset{OOH}{|}}{R-CH-R_1} \begin{array}{c} \nearrow R-CHO \\ \rightarrow \underset{\underset{OH}{|}}{R-CH-R_1} \\ \searrow \underset{\underset{O}{\|}}{R-C-R_1} \end{array}$$

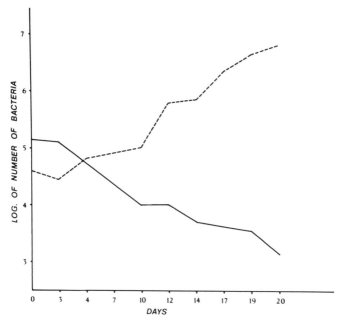

Figure 20. Number of Micrococci and Lactobacilli During the Fermentation of Fruskogorska Sausage. _____ micrococci, _ _ _ _ _ lactobacilli. After Stolić (1975).

catalyzed by various microorganisms. Alford et al. (1971) studied extensively the action of various bacteria on meat lipids and found that *Micrococcus freudenreichii* and *Pseudomonas ovalis (putida)* increased the peroxide value of fresh lard by factors of 14 and 8 respectively. Whereas *Mc. freudenreichii* promoted the formation of aldehydes (2-enals and 2,4-dienals), *Ps. fragi, Candida lipolytica,* and *Geotrichum candidum* also produced considerable amounts of methyl ketones (50 to 120 μmoles/10^4 μmoles of fat). On the other hand, microorganisms in contact with rancid lard showed considerable destruction of peroxides with or without formation of monocarbonyl compounds *(Aspergillus niger, Candida lipolytica).* There seems to be no relationship between the ability to decompose peroxides and the lipolytic action of these organisms. The formation of methyl ketones (C_5, C_6, C_7) has also been demonstrated with *Pediococcus cerevisiae* (Bothast et al., 1973). It is therefore not surprising that many processes taking place during the microbial action of meat curing do indeed affect the flavor profile of the product. Since the information on the exact flavor composition of such products is still not complete, it is difficult to provide precise environmental conditions and optimum microbial loads for each curing process.

The second group of microorganisms that appears during the curing process and dominates the microflora toward the final phases of the ripening of fermented sausages consists of lactic acid bacteria, which are also the least sensitive organisms to the bacteriostatic action of smoking (Urbaniak and Pezacki, 1975). It has already been pointed out that lactobacilli contribute to the ripening process by decreasing the pH to such values that retard the proliferation of various undesirable organisms (together with the protective effect of the salt) and thus promote the gelation process of meat proteins. In addition, the production of lactic acid constitutes a major part in the formation of the fer-

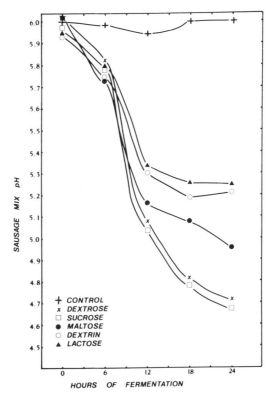

Figure 21. Rate of pH Reduction in Fermenting Sausage Containing Various Carbohydrates (1%) and a Starter Culture of *Pd. acidilactici*. After Acton et al. (1977).

mented taste (Fig. 21). Further, the pH drop exerts a certain control on the reductive reaction that takes place between the nitrate-reducing organisms and their substrate. Under conditions where the pH drop is prevented, nitrate reduction may become so vigorous as to damage the quality of the product by causing the so-called "nitrite burn" (Deibel et al., 1961). The participation of many lactic bacteria in the nitrate reduction process in meats has recently been suggested (Smith and Palumbo, 1978).

In the case of meat fermentation, one might also expect that the lactic microflora would be represented, at least during the initial phase, by heterofermentative *Leuconostoc* spp. as usually happens at the beginning of vegetable pickling (see Chapter 5). The presence of considerable amounts of volatile acids in cured sausages may be considered indicative of such a fermentation. Also, the fact that some manufacturers puncture the casing of sausages to permit escape of gases may suggest the occurrence of heterofermentative fermentation during this stage of the curing, although this has not yet been confirmed experimentally (Pederson, 1971). More recently, a greater role in the formation of sausage flavor was attributed to the heterofermentative lactics *Lb. brevis* and *Lb. buchneri*, the number of which seems to increase during the latter part of the ripening period (Urbaniak and Pezacki, 1975) (Table XXVIII).

The analogy between cheese ripening and the sausage fermentation has led many researchers to attempt using pure cultures for meat curing in order to obtain products with uniform characters of high quality and possibly also to accelerate the curing process itself. This was not accomplished before the

TABLE XXVIII
MEAN COMPOSITION OF DRY SAUSAGE

Dry matter	65%
Protein (% D.M.)	28%
Fat (% D.M.)	61%
Lactate (mmole/100 g D.M.)	28.3
Acetate (mmole/100 g D.M.)	2.4
Butyrate (μmole/100 g D.M.)	14.5
Propionate (μmole/100 g D.M.)	11.7
Carbonyl compounds (μmole/100 g D.M.)	334 ($\pm$ 64)
pH	4.87

SOURCE: After DeKetelaere et al. (1974).

pioneering studies of Niinivaara (1955) and Niven et al. (1958), who were probably the first to use bacterial curing starters on an industrial scale. Although certain workers attempted the use of a *Pseudomonas* culture in curing the brine of ham (McLean and Sulzbacher, 1959), commercial starter preparations concentrated on selected strains of a *Micrococcus* (Niinivaara, 1955) and a *Pediococcus* (Niven et al., 1955) or a mixed culture of both (Deibel et al., 1961; Everson et al., 1970). Others suggested the use of a certain strain of the gram-negative *Vibrio costicolus* (Petäjä et al., 1973; Petäjä, 1977). Initially, there were some doubts as to the beneficial effect of starter inoculations on flavor formation (Ingram, 1966); however, in recent years more and more industries adopted this procedure. Today, many preparations of concentrated (Rothchild and Olsen, 1971) or freeze-dried cultures (Rohde and Weckend, 1976; Gryzka, 1977) have become available — Accel®, Bactoferment-61®, Duplo-ferment-66®, to mention a few. The question of mixed versus monocultures has also risen in this area. While American manufacturers prefer the "fast ripening" process which leads to a sharp-sour taste (tangy), the Europeans seem to prefer the slower ripening procedure resulting in sausages with milder flavor. For this purpose, European manufacturers seem to prefer a micrococcal monoculture as starter (Coretti, 1977).

Since the use of nitrites for curing has been permitted in several countries, the use of *Micrococcus* no longer seems obligatory. On the other hand, *Pediococcus* starter cultures seem to be the organism of choice since the relatively weak homofermenter, contrary to *Lb. plantarum*, assures adequate pH reduction without running the risk of spoilage due to excessive souring. Other workers seem to disagree on this point (Reuter, 1972). On the other hand, the methyl ketone formation ascribed to the metabolic activities of *Pediococcus* may be important to the desired flavor buildup of fermented sausages (Bothast et al., 1973). None of the lactobacilli involved in these starters were found to contribute significantly to the level of biogenic amines in fermented sausage (Rice and Koehler, 1976). Additional information on the flavor components of cultured products would be helpful in the further development of specific starter cultures. The incorporation of heterofermentative lactics suggested earlier by Leistner (1963) should be reevaluated from the flavor viewpoint even though gas formation by such organisms would not be desirable. Sison and Pederson (1974), who studied the kinetics of cultured sausage production, concluded that with suitable starters and environmental conditions the ripening process may be terminated within one day (16 hours at 41°C and a subsequent 8 hours of smoking), thus achieving the long-sought-for acceleration of the sausage curing process. Abbreviation of the aging time of country style hams by the use of *Pediococcus* starters has also been reported (Bartholomew and Blumer, 1977).

In studying the role of microorganisms in the flavor formation of fermented meat products, a further analogy to cheese maturation might be attempted, i.e. the production of cultured sausages with meats practically devoid of nonstarter bacteria. This is a very difficult problem, especially with comminuted products. Attempts to suppress the natural microflora by the incorporation of antibiotics like streptomycin has met with only limited success since the starter organisms were also similarly retarded (Reuter, 1972). The use of delta gluconolactone (0.7%) instead of acid formation by lactobacilli was not successful either, due to color defects and certain aroma disadvantages (Coretti, 1971).

The microbiology of cured meats is by no means limited to bacterial organisms. The natural population of such meat products also comprises a number of yeasts and fungi. Szczepaniak et al. (1974) described yeasts isolated from all stages of *serwolatka* sausage manufacture (10^5 to 10^8/g dry weight). Akashi (1974) isolated two unidentified yeast strains from brines of meat curing and found them to be suitable for use as starter organisms because of their strong nitrate-reducing ability and low ammonia production. Sesma and Ramirez (1976) investigated the yeast microflora of Pamplona sausage (*solchichóns*) and found many oxidative yeasts such as *Debaryomyces* spp. The surface of sausages with natural or synthetic casings seem to be an excellent medium for the proliferation of such organisms during ripening and storage. In fact, many of the fermented meat products appear on the market with heavy layers of yeast growth. We still have little information about the role these organisms play, if any, in sausage flavor. Rossmanith and Leistner (1972) suggested the use of various yeasts (*Debaryomyces kloeckeri, D. cantarellii,* and *D. pfaffii*) in starters in order to impart a specific yeasty aroma desirable in many air dried sausages. This, however, has not yet been tested on a commercial scale. On the other hand, fungal organisms may develop considerably on sausage casings from environmental contamination. This is more frequent on cured products that undergo no smoking. Many species of the genera *Penicillium, Aspergillus, Cladosporium,* and others have been found on the surface of these products. Such mycelial foci may impart a mottled appearance because of the color shades of their respective spores which, of course, cannot be considered desirable for marketing. Recently, the hygienic problem of molded meat products has been raised since such mold has also been shown to produce certain mycotoxins. In order to overcome this problem, the immersion of cured hams and sausages into a suspension of spores of *Penicillium nalgiovensis* has been suggested (Liepe, 1972). This organism covers the entire surface of sausages within 4 to 5 days of incubation at 20 to 24°C with a colorless mycelium, thus avoiding any accidental contamination of other undesirable molds. *P. nalgiovensis* produces no mycotoxins. Improved aroma of ham covered with such growth has been claimed (Mintzlaff and Christ, 1973). The chemical entities of such mold cured meat products await further investigation.

FISH PRODUCTS AND DEFECTS

General Considerations

This is the appropriate section to say a few words about microbes and fish flavor. Fish cooked before rigor mortis sets in, i.e. within minutes after death takes place, have a "metallic" flavor (Jones, 1967). This, however, is rarely the case; fresh fish give a sweetish to slightly meaty character after conventional cooking. These flavors gradually disappear upon storage, even under ice or refrigeration. There seems to be a general enhancement of fish flavor owing to the rapid transformation of AMP to 5′IMP, which is followed by the degradation to inosine and hypoxanthine, the latter conferring some bitterness to the fish (Spinelli,

1971). All of these reactions are probably mediated by autolytic enzymes and differ with different species. Microbial effects will not begin until spoilage is nearly underway (Shewan, 1962; Shewan and Hobbs, 1967). We shall therefore not go into any further details about the chemical components of fish (for more information see Jones, 1967). However, some other aspects of fish flavor are worth a more detailed discussion.

Fermented Fish Products

Fish sauces and fish pastes are traditional products with large consumption in the Far East and southeastern countries of Asia, almost unknown in the western world (Van Veen, 1965). They appear under various names such as *nuoc-man, shottsuru, prahoc,* etc. Fish sauces are all prepared in a similar way. Small, noneviscerated fish are mixed with salt (1 part salt to between 1.5 and 3 parts of fish). The mixture is placed in concrete vats or other containers and is allowed to ferment for nine months at 30 to 35°C. The fish disintegrate, giving a brown liquid that is drawn off and allowed to mature for an additional three months. The mature liquid is similar in appearance to soy sauces, perhaps somewhat more pale with a characteristic sharp aroma compounded of ammoniacal, meaty, and cheesy notes (most of the nitrogen appears as free amino acids). There are good indications that this hydrolysis is carried out chiefly by fish enzymes. However, eviscerated fish produce only poor sauces. There is good reason to believe, therefore, that microbial activity has an important share in this fermentation. Inhibition of microbial activity by heat or antibiotics prevents the formation of the characteristic oriental flavor (Beddows et al., 1976). Early workers claimed that the specific aroma was due to a special strain of *Clostridium* sp.; however, this was not confirmed (Van Veen, 1965). Attempts to shorten the fermentation time by the addition of various proteolytic enzymes such as papain and increased temperatures also resulted in products with unsatisfactory flavor (Dougan and Howard, 1975). Fish sauces have a fairly constant flavor profile with regard to volatile fatty acids, mainly — acetic, propionic, butyric, and isovaleric acids. Mean values of 1:20 butyric to acetic acid ratios have been mentioned for different products (acout 400 mg acetic acid/100 ml of sauce). The meaty note of the sauce has not yet been analyzed but seems to be reminiscent of that of yeast or meat extracts. According to Saisithi and coworkers (1966), the distinct flavor of fish sauce is a blend of organic acids and a number of amines like histamine, glutamine, glucosamine, and trimethylamine. Low molecular weight volatiles, such as methyl ketones and other carbonyl compounds, have been suggested by others (Yurkowski and Bordeleau, 1965). The microbial contribution to the formation of these flavor components warrants more extensive investigations.

Attempts have also been made in Denmark to utilize fish not suitable for human consumption for the production of a so-called "fish cheese." In these experiments, eviscerated fish are minced, pasteurized, and stirred with water and whey powder or sucrose, and a starter culture of lactic acid bacteria (*Lb. helveticus* and *Lb. acidophilus*) is added together with potassium nitrate. A lactic fermentation sets in, reducing the pH to about 4.5. After 24 hours at 30°C, the curd is pressed and brined in 20 percent salt before packaging. A shelf life of several months is claimed (Herborg and Johansen, 1977). Although the technology of fish cheese is only in the beginning, a controlled lactic fermentation of fish may be of importance after several problems in the choice of fish and their preparation for the fermentation are solved. A suitable flavor for such a product will have to be secured before acceptance for consumption may be expected.

Flavor Defects in Fish

A very frequent off-flavor that makes fish practically unmarketable has been known for a long time as "muddy" or "earthy" fish. Both freshwater and saline lake fish suffer from such defects. Two types of microorganisms have been found to be involved in this defect: actinomycetes and blue-green algae. In both cases, extracellular products that accumulate in the water are absorbed by the fish during certain periods of the year when conditions become favorable for the proliferation of these organisms and the production of the earthy odor. Two main compounds seem to be involved in this off-flavor, one of which is geosmin (trans-1, 10-dimethyl-trans-9-decalol):

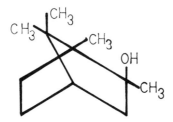

This compound has a very strong earthy smell and is recognizable at very low concentrations (even below 1 ppb). Another compound is 2-methyl isoborneol (2-exo-hydroxy-2-methyl bornane), which has a somewhat lacquerlike odor. After Rosen et al. (1970):

Interestingly, both types of compounds may be found in such different organisms like actinomycetes and blue-green algae. Among the actinomycetes, various species of the genus *Streptomyces* have been shown to produce geosmin and 2-methyl isoborneol. Among the blue-greens, various species of *Oscillatoria, Lyngbya,* and *Symploca* were shown to produce geosmin, while the formation of 2-methyl isoborneol seems to be more limited; so far, only *Lyngbya cryptovaginata* was found to synthesize these compounds. In saline lakes, most of these odoriferous substances are produced by the blue-greens; in fresh water, actinomycetes are also important (Tabachek and Yurkovski, 1976). The earthy odor of soil is probably due primarily to the activities of such actinomycetes (Gerber, 1967).

Another off-flavor of fish has been described during the initial phases of spoilage of chilled fish muscle. A musty, potatolike odor is produced by certain bacteria, e.g. *Pseudomonas perolens*. This was demonstrated with sterile fish muscle (Miller et al., 1973). The odor comprises several sulfur compounds as well as certain pyrazines, primarily 2-methoxy-3-isopropyl pyrazine, an extremely powerful odorant.

Fishiness, or a fishy off-flavor, may occur also in nonfish foods and is somehow related to the formation of trimethylamine. Fishiness in salted, stored beef was found to have been caused by a halophilic *Micrococcus* (Jensen, 1954). The reduction of trimethylamine oxide to trimethylamine by fish microflora is one of the main symptoms of fish spoilage. The occurrence of trimethylamine in nonfish foods may also be the result of non-biological reactions.

BIBLIOGRAPHY

Acton, J. C., Dick, R. L., and Morris, E. L.: Utilization of various carbohydrates in fermented sausage. *J Food Sci, 42:*174, 1977.

Akashi, A.: Effective strains for the curing of meat products. *Jpn J Dairy Sci, 23:*A151, 1974.

Alford, J. A., Smith, J. L., and Lilly, H. D.: Relationship of microbial activity to changes in lipids of foods. *J Appl Bacteriol, 34:*133, 1971.

Backus, J. N. and Deibel, R. H.: Nitrite burn in fermented sausage. *Appl Microbiol, 24:*405, 1972.

Bartholomew, D. T. and Blumer, T. N.: The use of a commercial *Pediococcus cerevisiae* starter culture in the production of country style hams. *J Food Sci, 42:*494, 1977.

Beddows, C. G., Ismail, M., and Steinkraus, K. H.: The use of bromelain in hydrolysis of mackerel and the investigation of fermented fish aroma. *J Food Technol, 11:*379, 1976.

Bothast, R. J., Kelly, R. E., and Graham, P. P.: Influence of bacteria on the carbonyl compounds of ground porcine muscle. *J Food Sci, 38:*75, 1973.

Buchanan, R. E. and Gibbons, N. E. (Eds.): *Bergey's Manual of Determinative Bacteriology*, 8th ed. Baltimore, Williams & Wilkins, 1974.

Cantoni, C., Molnar, M. R., Renon, P., and Giolitti, G.: Lipolytic micrococci in pork fat. *J Appl Bacteriol, 30:*190, 1967.

Coretti, K.: Rohwurstreifung und Fehlerzeugnisse bei der Rohwurstherstellung. Alzey, Verlag Rheiness, 1971.

————: Starterkulturen in der Fleischwirtschaft, *Fleischwirtsch, 57:*386, 1977.

Davey, C. L. and Gilbert, K. V.: Studies in meat tenderness. *J Food Sci, 31:*135, 1966.

Davidson, C. M., Mobbs, P., and Stubbs, J. M.: Some morphological and physiological properties of *Microbacterium thermosphactum*. *J Appl Bacteriol, 31:*551, 1968.

Debevere, J. M., Voets, J. P., de Schryver, F., and Houyghebaert, A.: Lipolytic activity of a *Micrococcus* sp. isolated from a starter culture in pork fat. *Lebensm Wissensch u Technol, 9:*160, 1976.

Deibel, R. H., Niven, C. F., and Wilson, G. D.: Microbiology of meat curing. III. Some microbiological and related technological aspects in the manufacture of fermented sausages. *Appl Microbiol, 9:*156, 1961..

DeKetelaere, A., Demeyer, D., Vandekerckhove, P., and Vervacke, I.: Stoichiometry of carbohydrate fermentation during dry sausage ripening. *J Food Sci, 39:*297, 1974.

Dougan, J. and Howard, G. E.: Some flavoring constituents of fermented fish sauces. *J Sci Food Agric, 26:*887, 1975.

Dwivedi, B. K.: Meat Flavor. *Crit Rev Food Technol, 5:*487, 1974.

Everson, C. W., Danner, W. E., and Hammes, P. A.: Bacterial starter cultures in sausage products. *J Agric Food Chem, 18:*570, 1970.

Fournaud, J.: La microbiologic due saucissonsec. *L'Alim et al Vie, 64:*82, 1976.

Gerber, N.: Geosmin, an earthy-smelling substance isolated from actinomycetes. *Biotechnol Bioeng, 9:*321, 1967.

Grau, R.: *Fleisch und Fleischwaren: Grundlagen und Fortschritte der Lebensmitteluntersuchung.* Bd. 7. Berlin, Verlag Hay's Erben, 1960, S.118.

Gryzka, A. J.: Meat fermentation. U.S. Pat. 4,013,797, 1977.

Harris, N. G., Strong, D. M., and Sunde, M. L.: Intestinal flora and chicken flavor. *J Food Sci, 33:*543, 1968.

Herborg, L. and Johansen, S.: Fish cheese: the preservation of minced fish by fermentation. *Proceedings of Conference on Handling, Processing, and Marketing of Tropical Fish.* London, Tropical Products Institute, 1977, p.253.

Herz, K. O. and Chang, S. S.: Meat flavor. *Adv Food Res, 18:*1, 1970.

Hornstein, I.: Flavor of red meats. In Schultz, H. V. (Ed.): *The Chemistry and Physiology of Flavors.* Westport, Connecticut, Avi, 1967.

Ingram, M.: Introductory remarks, with special reference to meat. Symposium on the Microbiology of Desirable Food Flavors. *J Appl Bacteriol, 29:*217, 1966.

Jensen, L. B.: *Microbiology of Meats,* 3d ed. Champaign, Illinois, Garrard, 1954.

Jones, N. R.: Fish Flavor. In Schultz, H. V. (ed.): *The Chemistry and Physiology of Flavors.* Westport, Connecticut, Avi, 1967.

Kitchell, A. G.: Micrococci and coagulase negative staphylococci in cured meats and meat products. *J Appl Bacteriol, 25:*416, 1962.

Koizumi, C. and Brown, W. D.: Formation of nitric oxide myoglobin by nicotinamide adenine dinucleotides and flavins. *J Food Sci, 36:*1105, 1971.

Kramlich, W. E., Pearson, A. M., and Tauber, F. W.: *Processed Meats.* Westport, Connecticut, Avi, 1973.

Leistner, L.: Rohwurst in der USA. *Arch Lebensmittelhyg, 14:*62, 1963.

Liepe, H. U.: Die praktische Anwendung von Starterkulturen bei Rohwurst. *Fleischwirtsch, 56:*178, 1976.

————: Luftgetrocknete Wurst mit Kultur-Schimmel. *Fleischwirtsch, 52:*967, 1972.

Lillard, D. A. and Ayres, J. C.: Flavor compounds in country cured hams. *Food Technol, 23:*251, 1969.

MacDougall, D. B., Mottram, D. S., and Rhodes, D. N.: Contribution of nitrite and nitrate to the color and flavor of cured meats. *J Sci Food Agric, 26:*1743, 1975.

Mast, M. G. and Stephens, J. F.: Effects of selected psychrophilic bacteria on the flavor of chicken breast meat. *Poult Sci, 51:*1256, 1972.

McLean, R. A. and Sulzbacher, W. L.: Production of flavor in cured meat by a bacterium. *Appl Microbiol, 7:*81, 1959.

Mihalyi, V. and Körmendy, L.: Changes in protein solubility and associated properties during the ripening of Hungarian dry sausages. *Food Technol, 27:*1398, 1967.

Miller, A., Scanlan, R. A., Lee, J. S., Libbey, L. M., and Morgan, M. E.: Volatile compounds produced in sterile fish muscle *(Sebastes melanops)* by *Pseudomonas perolens. Appl Microbiol, 25:*257, 1973.

Minor, L. J., Pearson, A. M., Dawson, L. E., and Schweigert, B. S.: Chicken flavor: the identification of some chemical components and the importance of sulfur compounds in the cooked volatile fraction. *J Food Sci, 30:*686, 1965.

Mintzlaff, H. J. and Christ, W.: *Penicillium nalgiovensis* als Starterkultur für Südtiroler Bauerspeck. *Fleischwirtsch, 53:*864, 1973.

Mirna, A. and Hofman, K.: Über den Verbleib von Nitrite in Fleischwaren. *Fleischwirtsch, 49:*1361, 1969.

Niinivaara, E. P.: The influence of pure cultures of bacteria on the maturing and reddening of raw sausage. *Acta Agric Fennica, 85:*95, 1955.

Niven, C. F., Deibel, R. H., and Wilson, G. D.: The use of pure culture starters in the manufacture of summer sausage. *Ann Meat* (Amer Meat Inst), *11:*22, 1955.

————: The AMIF sausage starter culture. *Found Bull* (Amer Meat Inst), *41 1958.*

Ockerman, H. W. and Cahil, V. R.: Microbiological growth and pH effects on bovine tissue inoculated with *Pseudomonas putrefaciens, Bacillus subtilis,* or *Leuconostoc mesenteroides. J Food Sci, 42:*141, 1977.

Patton, S., Barnes, I. J., and Evans, L. E.: n-Deca-2, 4-dienal, its origin from linoleate and flavor significance in fats. *J Am Oil Chem Soc, 36:*280, 1959.

Pederson, C. S.: Microbiology of Food Fermentations. Westport, Connecticut, Avi, 1971.

Petäjä, E.: Untersuchunger über die Verwendungsmöglichkeiten von Starterkulturen bei Bruhwürst. *Fleischwirtsch, 57:*109, 1977.

Petäjä, E., Laine, J. J., and Niinivaara, F. P.: Einfluss der Pökellakebakterien auf die Eigenschaften gepökelten Fleisches. *Fleischwirtsch, 53:*680, 1973.

Pezold, H. V.: Verderben und Vorrathaltung von Fette und Fettprodukte. In Schormüller, J. (Ed.): *Handbuch der Lebensmittelchemie,* vol IV. Berlin, Springer Verlag, 1969.

Piotrowski, E. G., Zaika, L. L., and Wasserman, A. E.: Studies on aroma of cured ham. *J Food Sci, 35:*321, 1970.

Pippen, E. L.: Poultry flavor. In Schultz, H. V. (Ed.): *Chemistry and Physiology of Flavors.* Westport, Connecticut, Avi, 1967.

Reuter, G.: Versuche zur Rohwurstreifung mit Laktobacillen und Mikrokokken-Starterkulturen. *Fleischwirtsch, 52:*465, 1972.

Rice, S. L. and Koehler, P. E.: Tyrosine and histidine decaroxylase activities of *Pediococcus cerevisiae* and Lactobacillus species and the production of tyramine in fermented sausages. *J Milk Food Technol, 39:*166, 1976.

Rohde, R. and Weckend, B.: Lyophil getrocknete Starterkulturen für die Fleischindustrie. *Fleisch, 30:*56, 1976.

Rosen, A. A., Mashni, C. I., and Safferman, R. S.: Recent developments in the chemistry of odor in water: the cause of earthy/musty odour. *Water Treat Exam, 19:*106, 1970.

Rossmanith, E. and Leistner, L.: Hefen als Starterkulturen für Rohwurste. *Mitteil Bund Fleischf, 38:*1705, 1972.

Rothchild, H. and Olsen, R. H.: Frozen *Pediococcus cerevisiae* concentrate with stabilizer and culture medium. U.S. Pat. 3,561,977, 1971.

Saisithi, P., Kasemsarn, B., Liston, J., and Dollar, A. M.: Microbiology and chemistry of fermented fish. *J Food Sci, 31:*105, 1966.

Scalan, R. A.: N-nitrosamines in foods. *Crit Rev Food Technol, 5:*357, 1975.

Sesma, B. and Ramirez, C.: Study of the microbial flora in ripening sausage. *Rev Agronoquim Technol Aliment, 16:*509, 1976.

Shank, J. L., Silliker, J. H., and Harper, R. H.: The effect of nitric oxide on bacteria. *Appl Microbiol, 10:*185, 1961.

Shewan, J. M.: The bacteriology of fresh and spoiling fish and some related chemical changes. In S. Hawthorn and Muil Leitch, J. (Eds.): *Recent Advances in Food Science,* vol. 1. London, Butterworth, 1962.

Shewan, J. M. and Hobbs, G.: The bacteriology of fish spoilage and preservation. *Prog Ind Microbiol, 6:*169, 1967.

Sison, E. C. and Pederson, C. S.: Fermentation of native smoked sausage. *Philipine Agriculturist, 58:*61, 1974.

Smith, J. L. and Palumbo, S. A.: Reduction of nitrate in a meat system by *Lactobacillus plantarum. J Appl Bacteriol, 45:*153, 1978.

Spinelli, J.: Biochemical basis of fish freshness. *Process Biochem, 6:*36, 1971.

Stolić, D. D.: Quantitative relationship between micrococci and lactobacilli during ripening of fermented sausages and factors influencing this relationship. *Acta Vet Beograd, 25:*91, 1975.

Szczepaniak, B., Urbaniak, Z., Spylczyn, K., and Pezacki, W.: Variations in yeasts of raw sausages. *Med Weterynaryjna, 30:*497, 1974.

Tabachek, J. L. and Yurkowski, M.: Isolation and identification of blue-green algae producing muddy odor metabolites, Geosmin, and 2-methylisoborneol, in saline lakes in Manitoba. *J Fisheries Res Board Can, 33:*25, 1976.

Tenfold faster flavor improvement of beef. *Food Process Marketing, 23:*83, 1963.

Urbaniak, L. and Pezacki, W.: Die Milschsäure bildende Rohwurst-Mikroflora und ihre technologisch bedingte Veränderung. *Fleischwissensch, 55:*229, 1975.

Van Veen, A. G.: Fermented and dried seafood products of South-East Asia. In Borgstrom, G. (Ed.): *Fish as Food.* New York, Acad Pr, 1965.

Yurkowski, M. and Bordeleau, M. A.: Carbonyl compounds in salted cod. *J Fisheries Res Board Can, 22:*27, 1965.

Chapter 5

VEGETABLE PRODUCTS

Pickles may be considered cooked.

—Babylonian *Talmud,* Hulin 111b.

PICKLED VEGETABLES

THE ACTION of lactic acid bacteria on plant material is utilized for the purpose of preserving nutritionally important ingredients as well as for producing different types of foodstuffs with a special appeal to the consumer. The technology of vegetable pickling has been described in great detail by Pederson (1971). Pickled cucumbers, olives, and sauerkraut are among the most popular foods with a very large consumption. Both the brine fermentation as well as the dry salt pickling methods are employed. From the microbiological point of view, this type of fermentation differs from that in the dairy industry not only in the nature of the raw materials but also in the fact that pickling involves a microaerophilic environment characterized by its high saline content. It is the salt content more than any other environmental factor that regulates the process of this lactic fermentation. Most of the microflora carried over from the field (spore formers and gram-negative bacteria with considerable pectolytic activity) are suppressed and give rise to lactic acid bacteria in a typical enrichment process. Although the lactics involved in vegetable pickling are obviously of the salt tolerant types *(Leuconostoc mesenteroides, Lactobacillus brevis, Lb. plantarum, Pediococcus cerevisiae),* the microbial changes that take place during the vegetable fermentation are governed by the process of acid formation itself, in addition to the salt concentration. Plant carbohydrates (reducing sugars, sucrose, and also pentoses) are fermented to lactic acid, which increases gradually to its highest levels. However, the tolerance of lactic acid bacteria to acidity + salt varies a great deal. During the early phases of the fermentation, the heterofermentative *Lc. mesenteroides* usually becomes

140

the predominant organism, vigorously fermenting sugars to lactic acid, acetic acid, ethanol, and carbon dioxide. When the acid concentration reaches 0.3 to 0.5 percent, according to the salt concentration (after an equilibrium between concentrations within the vegetable and its environment has been reached), these lactic organisms are inhibited, giving rise to the more acid tolerant bacteria (*Lb. plantarum, Lb. brevis,* and *Pd. cerevisiae*) which terminate the lactic fermentation. Acidities will usually reach 1 percent in brined pickles but may approach 2 percent in the dry salt fermentations like sauerkraut. This succession of lactic bacteria takes place in most cases of vegetable fermentation. The relative proportions are greatly affected by salt concentrations and temperature. At higher salt concentrations the growth of these organisms is considerably reduced, while at higher temperatures (over 30°C) there will be a faster fermentation (acid production) with a greatly reduced initial phase *(Lc. mesenteroides),* and a rapid overgrowth of the homofermentative lactics will take place (Pederson and Albury, 1954, 1969). The reason for the predominance of the *Leuconostoc* spp. in the initial phase is not entirely clear. It is possible that the lactobacilli become established only after the redox potential has been sufficiently reduced due to the oxygen displacement by CO_2 or owing to the fact that the *Leuconostoc* spp. have a shorter generation time or both. Stamer et al. (1971), who examined the behavior of these lactic bacteria in Seitz filtered cabbage juice, found that at 2.25 percent salt and pH 6.2, *Lc. mesenteroides* had the shortest lag phase and the shortest generation time of all cultures examined. On the other hand, *Pd. cerevisiae* was the most salt resistant and pH insensitive organism but would not grow in filtered juice in which *Lc. mesenteroides* had previously grown. This would explain why *Pediococcus* is not frequently encountered in normally conducted fermentations where *Lc. mesenteroides* usually predominates during the initial phase of the fermentation.

While the homofermentative lactic organisms are the principal acid producers, the heterofermentatives *Lc. mesenteroides* and *Lb. brevis* occupy an important position with regard to flavor formation, producing acetic acid, ethanol, and possibly also other flavor components, in addition to lactic acid. In considering the role of these organisms in the formation of the charac-

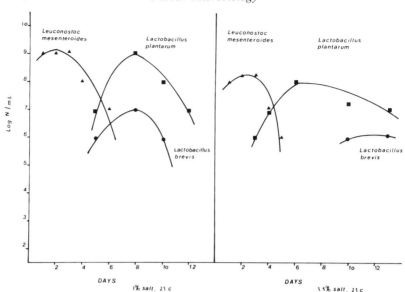

Figure 22. Effect of Salt Concentration on the Lactic Microflora in a Sauer-kraut Fermentation. Adapted from Pederson and Albury (1954).

teristic flavor of fermented plant material, we are confronted with a number of difficulties due to the complex nature of these processes. That lactic acid is not the only factor in the formation of the pickle flavor can be easily assessed by the fact that unfermented pickles prepared with similar amounts of such acid and salinity can be easily distinguished from fermented ones. However, during fermentation it is rather difficult to distinguish between the action of the microbial systems involved in the process and that of the enzymes that are present in the plant material. This difficulty is further aggravated by the fact that all vegetables employed in the lactic fermentation carry a very heavy load of microorganisms which do give way to the succession of lactic organisms but may also affect the final organoleptic qualities of the product. In order to establish the part played by the different groups of organisms, two pieces of information would be invaluable: (1) the approximate chemical composition of the flavor components of each type of the fermented products and (2) the chemical activity of each of these organisms involved in the lactic fermentation when these are investigated under pure culture conditions. Although this information is not easily obtained, recent studies have begun to concentrate upon these approaches.

An attempt to correlate the nonlactic acid components of pickles with the flavor quality of the fermented product has been made by Christensen and coworkers (1958), who analyzed the acetic acid composition of fermented vegetables. Using a setup that permitted pure culture fermentations, it was found that homofermentative lactics produced only minor amounts of acetic acid, independently of the sugar concentrations available, while with heterofermentatives the amount of acetic acid formed during the fermentation increased with sugar concentrations. With the heterofermentative *Lb. brevis*, acetic acid/lactic acid ratios of up to 4.9 were obtained (glucose). Although the contribution of such heterofermentative organisms to volatile acid formation in pickles may be assumed, quantitative data on acetic acid in the lactic fermentation of plant material are not abundant. Pederson et al. (1962) emphasize the importance of the velocity of the fermentation process to the organoleptic qualities of the product. In comparing the fermentation of Yugoslavian cabbage, it was found that whole cabbage kraut was superior to that of shredded kraut, the former developing a mellower flavor, while the latter showed a more pungent acid taste. This was probably due to the rapid fermentation that took place in the shredded cabbage leading to very high acidities. These results seem to be in accordance with earlier observations that suboptimal fermentation conditions (high salt concentrations, high temperatures, possibly also too high sugar concentrations) not favoring the activity of the heterofermenters in the initial stages yield pickles with poorer flavor qualities.

Do carbonyl compounds like acetaldehyde and diacetyl play a significant role in the formation of pickle flavor? Are the factors controlling diacetyl production in plant material (citric acid, pH, aeration, microflora) of similar importance to those in dairy products? If so, only during the initial stages of the fermentation, when oxygen is still available, diacetyl formation may take place. Indeed, Hardlička et al. (1964), who studied the formation of carbonyl compounds in kraut, found these compounds only during the first day of fermentation, declining thereafter.

A new approach to the study of the changes that take place during the vegetable fermentation was that of Vorbeck et al. (1963) and Pederson and coworkers (1964). These workers concentrated on the behavior of the lipid fraction during fermentation. It was found that in kraut, the amount of free fatty acid (FFA) increased considerably during the lactic fermentation. On the other hand, the amount of unsaponifiable matter and fatty acids of both acetone soluble and insoluble lipid fractions decreased during the fermentation (Table XXIX).

Similar results could be observed during the fermentation of Brussels sprouts. Among the free fatty acids, palmitic acid increased significantly during the fermentation (from 0.63 to 6.8% of total FA lipid in cabbage). The unsaturated C_{18} FA decreased, while the shorter chain fatty acids increased. The presence of longer chain saturated FAs in the nonesterified FA fraction of the fermented material has been attributed to changes in the unsaponifiable fraction and not only to the hydrolysis of lipids during fermentation. These longer chain FAs were notably absent from the FA of the acetone soluble and acetone insoluble lipids of the unfermented material. Since the role of plant lipolytic enzymes has not been taken into account, it is difficult to assess quantitatively the share of microbial activity. The chemical changes taking place during kraut fermentation have, however, not been discussed from the point of view of flavor formation and organoleptic evaluation. Pure culture studies with various lactic organisms may yield more information on the role played by these organisms.

TABLE XXIX
GROSS CHANGES IN THE VARIOUS LIPID FRACTIONS DURING
CABBAGE FERMENTATION

| Fraction | mg/200 g Dry Weight | | |
	Cabbage	Sauerkraut	% Change
Nonesterified fatty acids	86	649	+657
Acetone soluble lipid FAs	526	193	− 63
Acetone insoluble lipids FAs	978	504	− 49
Unsaponifiable matter	321	183	− 43

Source: After Vorbeck et al. (1963).

Information has also become available on the fate of the lipid fraction in the case of cucumber fermentations (Pederson et al., 1964). There was a general increase in the concentration of FFAs, neutral fat FAs and unsaponifiable material. Gas liquid chromatography of the methyl esters showed a sharp increase in linoleic (18 : 2) and linolenic (18 : 3) acids in the normal product, while in the case of "bloaters" an increase in oleic acid (18 : 1) took place. Furthermore, the disappearance of tridecenoic acid (13 : 1) from cucumbers and the appearance of caproic, caprylic, and capric acids were noted. A striking characteristic of the cucumber fermentation was the decrease of the phospholipid fraction, down to 10 percent of the original (Table XXX).

Higher values were observed in the analysis of bloated pickles. The change in the phospholipid fraction was not commensurate with the increase in FFAs, neutral fatty acids, and unsaponifiables. An active synthesis is therefore postulated to take place during the lactic fermentation. Incidentally, the breakdown of phospholipids during cucumber pickling has been noted earlier by Keil and Weyrauch (1937), who noted the accumulation of acetylcholine in fermented foods and ascribed it to the activity of *Bacterium acetylcholini*, which was later identified as a strain of *Lactobacillus plantarum* (Rowatt, 1948).

An interesting feature indirectly related to flavor has been recently ascribed to the metabolic activities of *Pediococcus* in the lactic fermentation. A Swiss group (Mayer et al., 1973) examined the sauerkraut fermentation from a nutritional point of view and noted the occurrence of considerable amounts of biogenic

TABLE XXX

CHANGES IN THE LIPID FRACTION DURING CUCUMBER FERMENTATION

Fraction	Cucumber*			Pickles*		
	Skin	Flesh	Seeds	Skin	Flesh	Seeds
Free fatty acids	392	300	455	1,890	1,557	3,838
Neutral FAs	430	233	188	667	623	1,454
Phospholipid FAs	1,203	642	1,232	50	72	75
Unsaponifiable matter	504	294	336	1,740	1,733	1,451

SOURCE: After Pederson et al. (1964).
* mg/100 g dry weight.

amines (histamine, tyramine) known to affect the blood circulation. While certain kraut samples contained up to 200 mg/kg histamine and 95 mg/kg tyramine, other samples were practically devoid of such amines. According to these authors, such amines are probably produced during the final phase of the fermentation after the pH has dropped below 3.8 and is probably related to *Pediococcus cerevisiae* known to ferment actively at high acid concentrations. Pasteurization of kraut prior to further pH reduction should eliminate this problem.

The role of *Lc. mesenteroides* in sauerkraut fermentation should also be considered from another point of view. Since this organism prefers the fructose moiety from sucrose, glucose accumulation takes place, resulting in the formation of the slime material, dextran. This process is apparently controlled by the prevailing salt concentration. At the commonly employed 2.25 percent salt concentration, visible slime formation does not occur, whereas at lower concentrations, excessive development of the *Leuconostoc* sp. will lead to considerable formation of slime and an unacceptable product. However, it is not unlikely that trace amounts of slime produced under normal fermentation conditions do favorably affect the texture and taste of shredded sauerkraut.

Pure culture fermentations of cucumbers have been attempted by several workers. A study of the organoleptic qualities of pure culture fermentations of cucumbers, accompanied by GLC analysis of high vacuum, low temperature distillates, has revealed that only *Lb. plantanum* cucumbers had a clean flavor, while both *Lb. brevis* and *Pd. cerevisiae* fermentations resulted in cucumbers with a somewhat bitter taste and a slight off-odor (Aurand et al., 1965).

Much of what is known about the chemical changes that take place during the pickling of cucumbers and sauerkraut may also apply to the fermentation of olives, although the latter fermentation differs from that of other plant fermentations in a number of features. The variability of olives (strain variations, stage of maturation), fat content, carbohydrate concentration, the occurrence of natural inhibitors to lactic organisms, e.g. the glycoside oleuropein, and type of processing (the lye treatment for removal of bitterness) may all affect the flavor profile of pickled olives. The olive fermentation is a lengthy one, and the microbial activities of the indigenous microflora cannot be excluded.

In order to gain more insight into the chemistry of olive pickling, pure culture fermentations have been attempted following a heat treatment to suppress the indigenous microflora (Etchells et al., 1966). Pure cultures of *Lb. plantarum*, *Lb. brevis*, *Pd. cerevisiae*, and *Lc. mesenteroides* were added to heat treated (74° × 3 minutes) Manzanillo olives — a variety known for their difficult fermentation. The heat treatment was effective not only in suppressing the competitive flora but also in making the olives more fermentable, i.e. facilitated the establishment of the inocula. *Lb. plantarum* was apparently most responsive to the heat treatment since it failed to become established in the unheated controls. On the other hand, *Lb. brevis* did not respond to this treatment, while *Pd. cerevisiae* and *Lc. mesenteroides* were intermediate. The nature of the inhibition of the lactic microflora in olive fermentation has been studied by various groups (Garrido-Fernandez and Vaughn, 1978). Etchells et al. (1966) were probably the first to describe the antilactic activity of the bitter principle of olives (oleuropein). Manzanillo olives contain large amounts of this compound (0.4% and over), which would explain the fermentation problems with this strain. However, Yuven and Henis (1970) suggested that most of the antibacterial activity was due to a degradation product of oleuropein. Since the antilactic activity of green olives could be overcome by a comparatively mild heat treatment (Etchells et al., 1966), the nature of the antilactic activity as well as the mode of its inactivation was questioned. A solution to this problem was advanced by the Corvallis group (Fleming et al., 1973) who suggested the following scheme: oleuropein is not the major principle of the antilactic activity but is transformed into a powerful antibacterial substance (the aglycone) by the enzymic activity (a β-glucosidase) of the green olives during the initial stages of the brining. Mild heat treatment that inactivates this enzyme is probably enough to prevent the antilactic activity in such fermentations. The aglycone is about ten times more inhibitory to lactic organisms than the glycoside. It is composed of elenolic acid bound through an ester linkage to 3,4-dihydroxyphenyl ethyl alcohol. Since the alcohol is not inhibitory, elenolic acid appears to be the inhibitory moiety of the aglycon. After Walter et al. (1975):

OLEUROPEIN   ELENOLIC ACID

Chemical analysis carried out by head space analysis has so far revealed acetaldehyde, methyl sulfide, and ethanol, as well as two unidentified peaks which were much higher than in nonheated, fermented controls. Lactic acid is considered the major flavor component of fermented olives (Fleming et al., 1969). Changes in the fat composition as a result of microbial activity have not yet been published.

Olive fermentations are usually accompanied by considerable growth of a variety of fermenting yeasts. Castanon and Inigo (1972) described a number of species (*Candida parapsilosis, C. utilis, Saccharomyces lactis,* as well as *Hansenula anomala, C. pelliculosa,* and *C. krusei*) according to the olive varieties fermented. That such yeasts may also contribute to acid formation has recently been suggested (Vaughn et al., 1976). The contribution of nonlactic organisms to the formation of flavor in the olive fermentation may constitute a new area of research in flavor microbiology. Yeasts present in the olive fermentation may also be involved in the formation of the antibacterial aglycone of oleuropein (Matterassi et al., 1975) although not affected by the toxic principle (Yuven and Henis, 1970).

Among the flavor defects known to occur in fermented vegetable products, we shall briefly mention the malodorous *zapatera* fermentation of olives. The first off-odor to appear in zapatera has been described as "cheesy," developing later into a foul, fecal stench. This is accompanied by a continuous loss of acidity. Delmouzos and coworkers (1953) have shown that at least part of the odor results from the volatile acids that develop in the brine; these include formic, propionic, butyric, valeric, caproic, and caprylic acids. In contrast, normal brines contain acetic, lactic, and sometimes succinic acids. Plastourgos and Vaughn (1957) believe that the propionic acid is produced by *Propionibacterium pentosaceum* and *Pr. zeae.* These organisms were shown to develop similar cheesiness *in vitro.* However, these authors emphasize that the propionic acid fermentation should be considered as only the first stage in the malodorous fermentation, to be followed by other anaerobic organisms. The environmental conditions leading to the zapatera fermentation have not yet been established with certainty.

ORIENTAL FOODS

The fermentation of vegetable products from leguminous plants (mainly soy beans) has not played any significant role in the Western world but was one of the most ancient processes in the Far East, the native area of soya. The reason for this difference may lay in the fact that the preservation of leguminous material requires very high salt concentrations which are not palatable to Western taste but are acceptable to Far Eastern peoples as a condiment for staple foods (such as rice) that are rather dull when consumed in quantities without fermented side dishes.

The fermentation of leguminous plant products, mainly soy beans, is completely different from that of the nitrogen poor plant material that is used for pickling. It is a much more complex process, both from the technological and microbiological point of view, and requires many months for completion (Yong and Wood, 1974). The dry soy beans, today mostly defatted, are first soaked overnight then cooked in an autoclave and mixed with wheat (roasted) and wheat bran (steamed) at roughly 1:1 proportions with a water content of 21 to 27 percent. The mixture is introduced into trays or small containers and inoculated with a culture of *Aspergillus oryzae* or *As. soyae.* After three days of incubation at 30°C with occasional stirring and cooling (optimal formation of proteolytic enzymes takes place at 20°C), the moldy (greenish spores) mixture becomes the *koji* ("bloom of mold") or starter for the next step. The koji is then soaked in a saturated salt solution of about equal volume (the *moromi* mash) which undergoes a slow fermentation (3 to 6 months at 15°C and gradual rise to 25°C). During this stage, proteins and carbohydrates are broken down by the koji enzymes; halotolerant bacteria, mainly *Pediococcus soyae (halophilus)*, begin a lactic fermentation decreasing the pH to below 5.0. During this time, osmotolerant yeasts *(Saccharomyces rouxi)* also develop and perform an alcoholic fermentation (2 to 6 months) which results in 2 to 2.5 percent ethanol. By the end of this period, the product (pH 4.7 to 4.8, 18% salt) is pressed, pasteurized, and marketed as *shoyu* or soy sauce. The moromi fermentation is regulated by its salt content and the fungal proteolytic and amylolytic enzymes that liberate fermentable material for the lactic and yeast organisms. Since this is a typical mixed flora fermentation, the organisms may vary a great deal. Yeasts such as *Zygosaccharomyces, Hansenula,* and *Torulopsis* have also been detected in this fermentation. More advanced industrial processes use pure cultures, e.g. *Lb. delbruekii* and *Hansenula* sp., to promote the fermentation. The final product is high in soluble proteins, peptides, and amino acids and has a dark brown color and a pleasant aroma.

Similar food fermentations are known in the Far East (Table XXXI). In most cases, the nature of the flavor of such fermented food products has so far not been analyzed, and little is known about the role played by the organisms in the formation of their respective flavors. We shall therefore review only what is known about shoyu, for which most of the information is available (Yokotsuka, 1960; Maga, 1973).

The chemical nature of soy sauce is a very complex one and is gradually formed during the various stages of its preparation and fermentation. During the soaking stage of the soy beans, a number of compounds are formed enzymatically, e.g. the mushroom flavor 1-octen-3-ol (Badenhop and Wilkens, 1969). However, the involvement of some microbial activity in this reaction should not be ruled out. In addition, several compounds that are known to be related to the unattractive soy aroma (ethyl vinyl ketone, lipid oxidation products such as 2 heptenal; 2,4 decadienal), as well as a series of phenolic compounds (bitter and

TABLE XXXI
IMPORTANT FAR EASTERN FERMENTED FOODSTUFFS

Name	Main Ingredients	Main Organisms	Duration	Taste
Miso	Rice, soybeans	*As. oryzae* (koji) Lactic bacteria Yeasts	1-3 months	paste (like peanut butter); 11-13% salt, 50% water
Shoyu	Soybeans, wheat	*As. oryzae* (koji) Lactic bacteria Yeasts	10-12 months	sauce; 18% salt
Natto	Soybeans	*Bacillus natto*	1-2 days	viscous; 50% water
Tempeh	Soybeans	*Rhisopus oligosporus*	20-24 hours	in banana leaves, sliced and fried
Sufu	Soybean milk, heat coagulation	*Mucor* sp.	3-7 days	salt brining with rice wine; aging 40-60 days 74% water, cheeselike

SOURCE: Compiled from Hesseltine and Wang (1972).

astringent), may all contribute to the flavor profile of the soy sauce. On the other hand, wheat and wheat bran incorporated into the koji preparation contribute a series of guaiacol compounds: vanillin, vanillic acid, ferulic acid, and 4-ethyl guaiacol, which are probably derived from lignin and the breakdown of various glycosides during the heat treatment.

VANILLIN   FERULIC ACID   VANILLIC ACID   4 - ETHYLGUAIACOL

During the koji fermentation, these compounds undergo various transformations. While vanillin and ferulic acid almost disappear, there is a gradual increase in the concentration of vanillic acid and 4-ethyl guaiacol, reaching a maximum before sporulation of the mold takes place. These compounds are probably related to the characteristic flavor of soy sauce. High quality shoyu contains 0.5 to 2.0 ppm 4-ethyl guaiacol. Less is known about the flavor evolution during the moromi fermentation. In addition to the production of lactic acid and ethanol by their respective fermentations, a remarkable decrease in the concentrations of malic and citric acids has been observed. The appearance of 4-ethyl phenol (phenolic), 2-phenyl ethanol (β-phenyl ethyl alcohol, floral-rosy), and tyrosol (bitter), as well as certain sulfur-containing compounds, through the action of various yeasts, may also be important to the characteristic flavor profile of soy sauce. In a more recent series of papers, Nunomura and coworkers (1976, 1978) describe a large number of flavor compounds of soy sauce as detected by GLC-MS analysis. Among these, certain furanones as well as pyrazines have been described. The origin of these compounds is still not clear, although it is suggested that most of the pyrazines are derived from the heating step and not the fermentation process.

TROPICAL FOODS

Many other plant materials are subjected to a process of fermentation in the curing phase prior to further processing. These processes are directed toward (1) the liberation of the main food ingredient from their accompanying envelopes and (2) the transformation of the plant material into compounds that impart specific flavor characteristics. The fermentation step is not a true fermentation in the microbial sense in all cases. In the case of tea production, the fermentation of tea leaves is actually an enzymatic process mediated by the plant enzymes that perform oxidative reactions which lead to the darkening of the leaves and the development of tea aroma. In the case of coffee, the coffee cherries contain two beans that are embedded in a thin, membraneous, parchmentlike endocarp which is surrounded by a yellowish pulp. The fermentation of these coffee cherries involves the digestion of this pulpy material which frees the seeds with their endocarp to be mechanically removed later. Most of the pulpy material is of pectic nature and, not surprisingly, the degradation of this material is carried out by pectolytic bacteria like *Erwinia* sp. and other enterobacterial organisms (Frank et al., 1965).

The fermentation of cacao beans is carried out with a similar purpose but has a somewhat different nature due to the chemical composition of the cacao fruit. About 30 to 40 cacao beans are contained in the cacao fruit, enmeshed in a white, sweet pulp (Fig. 23). The cacao fruit is cut open, and the pulpy beans are fermented to remove the pulp. Since this is a sweet material (about 10% sugar), the fermentation that sets in has an alcoholic nature followed by the production of acetic acid by acetic acid bacteria. The fermentation depends a great deal on the specific technology involved in this step. In locations where the pulpy beans are tightly packed, lactic acid bacteria also play a role in the fermentation. The seeds become dark brown and lose their viability. After drying, the outer shell may be readily removed. According to Forsyth and Quesnel (1963), the death of the seeds is caused by the acids and ethanol as well as the heat that evolves during the fermentation. For a good cacao fermentation, it is necessary that a low temperature step precedes the second, high temperature phase. Experiments have shown that dead beans will not develop the typical flavor formed during the initial

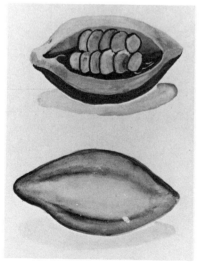

Figure 23. The Cacao Fruit (drawing).

stages of the germination, and neither will a characteristic cacao flavor develop in the absence of the second stage (several days at 50°C) (Wadworth and Howat, 1954).

Although the fermentation of these tropical fruits is an important step in the manufacture of coffee and cacao beans, little has been done so far in order to evaluate the possible contribution of the different groups of microorganisms to the flavor profile of their respective beverages.

BIBLIOGRAPHY

Aurand, L. W., Singleton, J. A., Bell, T. A., and Etchells, J. L.: Identification of volatile constituents from pure-culture fermentation of brined cucumbers. *J Food Sci, 30:*288, 1965.

Badenhop, A. F. and Wilkens, W. F.: The formation of 1-octen-3-ol in soybeans during soaking. *J Amer Oil Chem Soc, 46:*179, 1969.

Castanon, M. and Inigo, B.: Yeast occurrence and significance in the fermentation of spanish-green olives. *Microbiol Esp, 25:*195, 1972.

Christensen, M. D., Albury, M. N., and Pederson, C. S.: Variation in the acetic-acid lactic acid ratio among the lactic acid bacteria. *Appl Microbiol, 6:*316, 1958.

Delmouzos, J. G., Stadtman, F. H., and Vaugn, R. H.: Acid constituents of Zapatera of olives. *J Agr Food Chem, 1:*333, 1953.

Etchells, J. L., Borg, A. F., Kittel, I. D., Bell, T. A., and Fleming, H. P.: Pure culture fermentation of green olives. *Appl Microbiol, 14:*1027, 1966.

Fleming, H. P., Walter, W. M., and Etchells, J. L.: Isolation of a bacterial inhibitor from green olives. *Appl Microbiol, 18:*856, 1969.

————: Antimicrobial properties of oleuropein and products of its hydrolysis from green olives. *Appl Microbiol, 26:*777, 1973.

Forsyth, W. G. C. and Quesnel, W. C.: The mechanism of cacao curing. *Adv Enzymol, 25:*457, 1963.

Frank, M. A., Lum, N. A., and Dela Cruz, A. S.: Bacteria responsible for mucilage-layer decomposition in Kona coffee cherries. *Appl Microbiol, 13:*201, 1965.

Garrido-Fernandez, A. and Vaughn, R. H.: Utilization of oleuropein by microorganisms associated with olive fermentations. *Can J Microbiol, 24:*680, 1978.

Hardlička, J., Čurda, D. and Pavelka, J.: A study of the volatile carbonyl compounds during the fermentation of cabbage. *Sci Papers Inst Chem Technol, E15:*51, 1967.

Hesseltine, C. W. and Wang, H. L.: Fermented soybean food products. In Smith, A. K and Circle, S. J. (Eds.): *Soybeans: Chemistry and Technology.* Westport, Connecticut, Avi, 1972.

Keil, W. and Weyrauch, L.: Zur Pharmakologie und Chemie vergorener Nahrungsmittel. *Zentrbl Bakteriol, 97:*90, 1937.

Maga, J. A.: A review of flavor investigations associated with soy products, raw soybeans, defatted flakes and flours, and isolates. *J Agric Food Chem, 21:*864, 1973.

Matterassi, R., Miclans, N., and Pelagetti, O.: Hydrolysis of oleuropein in yeasts. *Ann Inst Sperim Elaiotech, 5:*53, 1975.

Mayer, K., Pause, G., and Vetsch, U.: Bildung biogener Amine während der Sauerkrautgärung. *Indust Obst Gemüseverwert, 58:*307, 1973.

Nunomura, N., Sasaki, M., Asao, Y., and Yokotsuka, T.: Isolation and identification of 4-hydroxy-2(or 5)-ethyl-5(or 2)-methyl-3(24)-furanone, as a flavor component in *shoyu* (soy sauce). *Agric Biol Chem, 40:*491, 1976.

————: *Shoyu* (soy sauce) volatile flavor components: basic fraction. *Agric Biol Chem, 42:*2123, 1978.

Pederson, C. A. and Albury, M. N.: The sauerkraut fermentation. *NY State Agric Exp Sta Bull, 824,* 1969.

Pederson, C. S.: *Microbiology of Food Fermentations.* Avi, Westport, Connecticut, 1971.

Pederson, C. S. and Albury, M. N.: The influence of salt and temperature on the microflora of sauerkraut fermentation. *Food Technol, 8:*1, 1954.

Pederson, C. S., Mattick, L. R., Lee, F. A. and Butts, R. M.: Lipid alterations during the fermentation of dill pickles. *Appl Microbiol, 12:*513, 1964.

Pederson, C. S., Niketic, G., and Albury, M. N.: Fermentation of Yugoslavian pickled cabbage. *Appl Microbiol, 10:*86, 1962.

Plastourgos, S. and Vaughn, R. H.: Species of propionibacterium associated with Zapatera spoilage of olives. *Appl Microbiol, 5:*267, 1957.

Rowatt, E.: The relation of pantothenic acid to acetylcholine formation by a strain of *Lactobacillus plantarum. J Gen Microbiol, 2:*25, 1948.

Stamer, J. R., Stoyla, B. O., and Dunckel, B. A.: Growth rates and fermentation patterns of lactic acid bacteria associated with the sauerkraut fermentation. *J Milk Food Technol, 34:*521, 1971.

Vaughn, R. H., Joe, T., Crampton, V. M., Lieb, B., and Patel, I. B.: Acid forming yeasts and the fermentation of olives. *Abst Am Soc Microbiol, 76:*188, 1976.

Vorbeck, M. L., Albury, M. N., Mattick, L. R., Lee, F. A., and Pederson, C. S.: Lipid alterations during the fermentation of vegetables by the lactic acid bacteria. *J Food Sci, 28:*495, 1963.

Wadworth, R. V. and Howat, G. R.: Cocoa fermentation. *Nature, 174:*392, 1954.

Walter, W. M., Fleming, H. P., and Etchells, J. L.: Preparation of antimicrobial compounds by hydrolysis of oleuropein from green olives. *Appl Microbiol, 26:*773, 1973.

Yokotsuka, T.: Aroma and flavor of Japanese soy sauce. *Adv Food Res, 10:*75, 1960.

Yong, F. M. and Wood, B. J. B.: Microbiology and biochemistry of soy sauce fermentation. *Adv Appl Microbiol, 17:*157, 1974.

Yuven, B. and Henis, Y.: Studies of the antimicrobial activity of olive phenolic compounds. *J Appl Bacteriol, 33:*721, 1970.

Chapter 6

FLAVOR IN THE PANARY FERMENTATION

. . . and bread to strengthen man's heart.

PSALMS 104:15

SOME BREAD TECHNOLOGY

OUR DAILY BREAD should be considered as one of the earliest and more important inventions of mankind. The exact origin of this invention is unknown to us but was probably related to the observation, during the early days of our civilization, that cereal grains may be ground up and mixed with water. The baking of such a mixture of flour and water thus constituted the early bread and would approach in its organoleptic properties that of the Jewish *matza* or the Indian *chapatties*. The leavening of such a mixture and the formation of the raised dough would constitute the second major breakthrough in the invention of bread. The raising power in the dough was undoubtedly attributed to some demonic force. It took several thousands of years until the microbial nature of leavening could be established. According to current concepts of fermentation, it is very likely that early leavening techniques in the hands of the housewife or the skilled baker followed a dichotomous line. Those who preferred a strong leavening action, as noted by the large volume of the product, supported the propagation of a yeast fermentation accompanied also by bacterial activity. Others who preferred the more acid leavening of bread thus led to an enrichment of lactic acid forming bacteria in the presence of varying amounts of yeasts. The prolonged kneading of the flour/water mixture and the use of part of the leavening from an earlier and successful trial thus ensured the outcome of the panary fermentation initiated by a spontaneous process. The transition from unleavened to leavened bread has been echoed for many generations in the form of religious rituals, such as the Passover rules of the Hebrews, forbidding leavened bread dur-

156

ing the seven days of Passover or the prohibition of leavened bread on the altar in the Holy Temple of Jerusalem. A practical point, however, is how to define what may be considered as unleavened bread. The Sages of the Babylonian Talmud (Pessachim 46a) decided that leavening starts within the time required for walking one mile, i.e. approximately eighteen minutes after the flour was brought in contact with water. Incidentally, this time period is very close to the generation time of a rapidly dividing bacterium, for example, *Escherichia coli.*

The third major step in the evolution of modern bread making is considered to be the introduction of compressed yeast, a pure culture of a strain of *Saccharomyces cerevisiae,* propagated for many generations under conditions that were not very aseptic before the culture was harvested and compressed into packages of yeasts (27 to 29% solids). The introduction of pressed yeast made work in the bakery much simpler and better suited most of the mechanized processes that evolved in the baking industry. In addition to the rapid dough rising required, sanitary considerations also made the use of the old leavening procedure less attractive. It is clear, however, that the use of compressed yeast changed the organoleptic features of bread toward a more voluminous but less acid bread (pH 5.3 to 5.7). On the other hand, leavening of the more sour type *(Sauerteig)* continues to be employed in a number of special breads (French sour bread, rye bread, and pumpernickel), resulting in bread with considerable acidity (pH 3.8 to 4.5).

The microbiology of bread centers, therefore, on two groups of microorganisms: (1) *S. cerevisiae* as main component of compressed yeast and (2) the bacterial flora developing in the dough — various types and genera of the lactic acid bacteria as well as the gram-negative nonspore-forming microbes that abound in flour, in addition to the ubiquitous spore formers. While microbes occupy a very important position in the buildup of the flavor profile of bread, this is not exclusive. The flavor of bread is very complex and depends on a large number of factors: the nature and quality of the ingredients (type of flour, amount of salt and sugar), the character of dough preparation, the fermentation conditions and finally, the type of baking employed. Since several excellent handbooks on the technology of baked products are available, e.g. (Matz, 1972), we shall mention briefly only the main types of modern bread making.

1. THE "STRAIGHT-DOUGH" METHOD. This involves only one mixing of all ingredients and a fermentation period of two to four hours before the dough is divided (the makeup) and molded. The quickest of the straight-dough methods would be the "no-time dough," where no fermentation time is allowed after mixing the ingredients. The dough is immediately molded and processed. This method is limited for emergency baking only, due to flavor considerations.

2. THE "SPONGE AND DOUGH" METHOD. Here, two distinct operations are employed. First, about 60 percent of the flour and water with all the yeast is mixed and allowed to ferment for three to five hours; after that, the "sponge" is returned to the mixer and mixed with the remaining ingredients (salt, shortening, etc.). After an additional fermentation period, the dough is molded and processed. Bread produced by this method usually has a greater volume and a more desirable texture and flavor.

3. THE PRE-FERMENT METHOD. This is a comparatively recent technique developed for use in continuous bread making. The pre-ferment generally consists of a mixture of yeasts, sugar, salts or milk, and water. The suspension is held for a number of hours (usually six at 100°F), with agitation before it is mixed with the flour.

An important modification of these methods is the Chorleywood procedure, which was developed in Britain but gained popularity in other countries also. According to this procedure, the dough is worked up in a fairly short time (minutes) by vigorous mechanical mixing (40 joules per g dough) without any pre-ferment but with higher loads of compressed yeasts. There is no fermentation holdup prior to molding.

From what has been said so far, there are three stages where intense microbial activity takes place: (1) the dough-rising stage or its equivalent in the pre-ferment brew (but absent in the Chorleywood process); (2) the proofing stage, i.e. the fermentation period after the makeup; and (3) the initial phase of the baking stage before temperatures become inhibitory to microbial activity, also called the "oven spring" period.

In order to assess the leavening reactions in dough, it is advisable to examine a typical formulation of bread (Table XXXII). From the microbiological point of view, the high concentration of yeast (which in the Chorleywood process may reach up to 6% on flour) is indicative of the importance of this organism in bread making. Contrary to many other food production processes that involve microbial action, in bread making, the role played by yeast is primarily mechanical, i.e. the production of carbon dioxide which extends and stretches the spaces between the gluten network. A few additional data regarding this reaction seem appropriate. The production of CO_2 under the fermentative conditions that prevail in the dough follows the well-known Gay-Lussac equation (Glucose $\rightarrow 2CO_2 + 2$ ethanol) to the extent of at least 95 percent of the total breakdown of the sugar. In quantitative terms, 1 g of yeast solids can ferment up to 3.35 g of glucose per hour, or 1 g of compressed yeast may ferment close to 1 g of sugar/hour and produce about 0.5 g of CO_2 per hour if adequate amounts of fermentable sugar are present. Since CO_2 is fairly soluble in water (over thirty-five times more soluble than oxygen), much of the CO_2 formed will concentrate in the water phase of the dough. Heating and baking of bread will thus promote the further liberation of the gas and lead to increased volume (up to four times that of the initial dough volume). Temperature is a very important factor in the leavening reaction; while the optimum for yeast growth is

TABLE XXXII
FORMULATION OF SOME TYPICAL BREADS (IN PARTS)

Ingredients	Vienna	Whole Wheat
Flour	100	100
Water	60	62
Salt	1.75	2
Sugar	1.75	3.5
Milk (MSNF)		2
Shortening	3	3
Malt syrup		0.5
Yeast (compressed)	1.75	3

Source: After Matz (1973).

around 30°C, the optimum for the fermentation of bakers' yeasts is closer to 38°C. Sponges and pre-ferments are usually maintained at temperatures below 30°C but are permitted to rise above that during the fermentation. On the other hand, proof boxes are generally held at 40°C or even higher in order to reach maximum gas production as quickly as possible (Reed and Peppler, 1973).

The substrate for the gassing action of yeast leavening is of course any fermentable sugar present in the flour, added to it, or produced from flour starch. Flour contains at least 0.4 percent fermentable sugars (Tanaka and Sato, 1969). If no sugars are added, the bulk of gas produced is derived from maltose formed during the leavening by certain amylolytic enzymes, the nature of which still remains to be clarified. According to Davis and Stephens (1954), about 5 percent of the starch in the dough is hydrolyzed and fermented. However, maltose requires a special enzyme (an α-glucosidase, maltase) which splits it into glucose before it can be fermented. This is apparently an inductive enzyme which requires some time to be formed (Suomalainen et al., 1972). Some workers suggested the use of a maltose syrup during the production phase of bakers' yeast in order to ensure high maltase activity in the dough (Drazhner et al., 1973).

It is very difficult to assess the leavening that takes place during the "oven spring" period. If oven temperatures are maintained in the range of 200 to 230°C during the approximately twenty minute baking cycle, the temperature in the center of the loaf will not exceed 93°C (Jackel, 1969), but the time left before yeast enzymes are inactivated (above 55°C) cannot be very long.

YEASTS AND BACTERIA IN BREAD MAKING

In spite of the mechanical and rheological attributes of the leavening, yeasts play a much more important role in bread production. Bread may be raised by chemical means (the so-called baking powder, a bicarbonate with a suitable acidulant) which may also proceed slowly if delta gluconolactone is employed. However, in most of the products in which yeasts are preferred, it is because of the specific flavor contribution to the baked product. In fact, when bread is leavened chemically, without any fermentation, the resulting flavor is substantially different from that of the typical bread flavor. If so, what are the nonstructural aspects, i.e. the chemical aspects, of yeast flavor?

We have good reason to believe that the yeast flavor is not the result of the metabolic activities of yeasts but the sensation caused by the consumption of the biomass per se. The removal of the volatiles of any product that underwent a yeast fermentation does not affect the basic yeasty character (Heiman and Drawert, 1968). Bakers' yeast consists of about 80 percent (dry weight) high molecular materials, including cell wall polymers, proteins, and nucleic acids; and about 20 percent low molecular materials, such as free amino acids, organic acids, and vitamins. Höhn and Solms (1975) have examined the relationship of the yeasty character with these fractions of bakers' yeast and found the yeast sensation to be related to the low molecular weight material. Thiamine and thiamindiphosphate seem to be responsible for the somewhat unpleasant and medicinal flavor of pure yeasts. According to Kurkela and Uutela (1972), the threshold values for thiamine sensation are the following:

Taste	0.3 to 20 ppm
Odor	0.00245 to 0.0098 ppm

with the values for thiamindiphosphate being about twenty times higher. It is interesting, therefore, to compare the thiamine content of various organisms (Table XXXIII).

As can be seen from Table XXXIII, the contribution of yeasts to the yeasty flavor of foods via its thiamine content is quite pronounced when compared to other plant or animal materials. In the case of baked products where yeasts are employed at high concentrations (up to 10% in the case of sweet doughs), high values of thiamine and a strong yeasty flavor may be expected.

THIAMINE

The thiamin molecule is composed of the pyrimidine moiety (4-amino-2,5-dimethylpyrimidine) and the thiazol moiety (4-

TABLE XXXIII
THE THIAMINE CONTENT OF VARIOUS FOODS

	mg/100 g D.W.
Torula yeasts	0.53
S. cerevisiae (Baker's)	3.0
S. carlsbergensis (Brewer's)	10.0
Liver	0.25
Potato	0.15
Beans	0.09

SOURCE: After Höhn and Solms (1975).

methyl-5-hydroxy ethyl thiazole). It seems as if the thiazol is the moiety responsible for the yeasty flavor, while the pyrimidine fraction, being 22.5 times less potent, enhances the typical flavor if attached to the thiazol moiety. Interestingly, a mixture of amino acids and sugars with thiamin yields a strong, meatlike flavor on heating (Bidmead, 1968).

Compressed yeasts contain about 10^9 to 10^{10} cells per gram. However, such yeasts as used in the baking process are not pure but contain large amounts of various bacteria which accompany the yeast propagation in the factory. White (1954) claims that commercial, compressed yeasts may tolerate up to 10^7 bacteria per gram. This number is likely to increase during dough making as a result of reproduction and infection from other ingredients as well as bakery utensils. Since these organisms amount roughly to only 1 percent of the number of yeast cells, there is a tendency of many authors to overlook their importance. However, the metabolic activities of such bacteria (mainly lactobacilli and coliforms) are so intense that their contribution to the formation of flavor in baked products must not be neglected. On the contrary, this accidental bacterial population constitutes an important part of bread flavor. In fact, according to Carlin (1958) bread prepared from dough raised with a pure culture of bakers' yeast obtained a flavor score of only 6.0 in comparison to 7.2 of bread leavened by commercial yeast. However, most authors dealing with the evolution of flavor in baked products do not distinguish between the contribution of yeasts and other organisms to the formation of taste and odor.

The ratio between yeasts and bacteria will, of course, be different in the case of a *Sauerteig* type dough. There will be about 100 to 1000 times more bacteria than yeasts (Spicher and Scholl-hammer, 1977). Sourdoughs practically devoid of yeast are known in India and the Philippines (*Idli* and *Puto*) in which the leavening is almost entirely due to the action of CO_2 forming heterofermenting lactics, usually *Leuconostoc mesenteroides* (Pederson, 1973). An interesting association of microbial forms in the San Francisco sourdough was described by Sugihara and associates (1971). These workers found no yeasts of the *S. cerevisiae* type but rather the maltose-negative yeast *S. exiguus* (*Torulopsis homii*), as well as the galactose-negative *S. inusitatus*. Here the number of bacteria in the dough was 30 to 100 times higher than the number of yeasts. A similar microbial association (*S. exiguus* and lactic acid bacteria) was also described in the Sour Dough of the Milanese *panettone* (Galli and Ottogalli, 1973).

THE FLAVOR OF BREAD

The problem of bread flavor differs from that of other fermented products in our interest to present to the consumer a product that can instantly be recognized as fresh and that retains its freshness as long as possible (Otterbacher, 1959).

For a long time studies on the flavor of bread were delayed by the fact that the typical appealing and pleasant flavor, sometimes also showing a so-called "interesting note," is far from stable but declines and disappears altogether with the staling process of the product. Attempts to chemically define the "good" flavor of bread have led to the analysis and identification of many compounds but, like in many other cases, we still do not know the exact nature of the desirable bread flavor. This prevents us from employing synthetic formulas for the enrichment of flavor poor breads we meet with today in many industrialized countries.

In analogy to the formation of meat flavor, it is the heating (baking) that creates the ultimate bread flavor; unbaked doughs cause a totally different sensation. Hence, the thermal process transforms many precursors, only part of which derived from microbial activity, into flavor constituents. Other ingredients comprise salts, sugar, flour, shortening, syrups, etc., each exerting their flavor attributes.

During the baking process, two events that ultimately affect bread flavor take place. First, a large number of volatiles are lost. Ethyl alcohol, the major product of the yeast fermentation along with CO_2, dissipates in the oven, leaving only small amounts in the product (about 4 to 6 g per kg bread). On the other hand, many products that considerably affect flavor are formed as a result of the condensation of certain products during the heating. Maga (1974), in his exhaustive review on bread flavor, lists close to 200 chemical entities involved in the evolution of bread flavor. Because of the complexity of the material, it is hard to assess the origin of all these compounds. In limiting ourselves to the flavor components contributed by microbial activity, one could surmise the origin of such compounds by following their quantitative behavior during the fermentation step of dough making. In fact, a number of workers examined this point.

As can be seen from Tables XXXIV through XXXVII, an increase in the concentration of organic acids, keto acids, and aldehydes can be observed during dough rising or in the preferment. However, there is a general decrease in the concentra-

TABLE XXXIV
THE ORGANIC ACID CONTENT OF FLOUR AND DOUGH*

	Flour	Dough
Acetic	0.05	0.36
Lactic	0.1	1.22
Glycolic (hydroxy acetic)	0.02	0.14
Formic	0.1	0.11
Pyruvic	0.01	0.18

SOURCE: Markova (1970).
* % dry matter.

TABLE XXXV
KETO ACID PRODUCTION IN PRE-FERMENTS*

Time hours	Pyruvic	α-ketoisovaleric	α-ketoglutaric
0	0.11	0	0
3	18.0	0.80	0.22
6	17.0	0.78	0.25

SOURCE: Cole et al. (1962).
* mmole/liter of pre-ferment.

TABLE XXXVI

CHANGES IN THE LEVEL OF SOME ALDEHYDES DURING FERMENTATION*

Compound	Before Fermentation	After Fermentation	After Proofing
Furfural	1.12	1.70	1.76
Benzaldehyde	2.28	4.80	4.88
Oxybenzaldehyde	0.80	1.07	4.25
Phenyl acetaldehyde	0.15	0.06	0.13

SOURCE: After Roiter and Borovikova (1971).
* mg/100 g.

TABLE XXXVII

CHANGES IN THE CONTENT OF SOME AMINO ACIDS
DURING FERMENTATION*

Amino Acid	0 Hours	6 Hours
Aspartic acid	0.55	0.05
Threonine and serine	1.08	0.13
Glutamic acid	0.58	0.11
Proline	0.08	0.10
Glycine	0.17	0.26
Alanine	1.08	0.07

SOURCE: After Morrimoto (1966).
* μg/g (dry weight).

tion of free amino acids, with the exception of proline and glycine, at least during the early hours of fermentation. In pre-ferments, a large variety of organic acids were detected. In addition to the aforementioned, isobutyric, valeric, crotonic, isocaproic, heptylic, caprylic, perlargonic, capric, lauric, myristic, and palmitic acids were also found (Hunter et al., 1961). Among the nonacid fractions of dough flavor constituents, the following alcohols were found in pre-ferments: n-propyl, isobutyl, amyl, and isoamyl, as well as 2,3-butanediol and 2-phenylethyl (Smith and Coffman, 1960). Kohn et al. (1961) examined the carbonyl compounds from straight dough prepared with compressed yeast and with a special yeast preparation with a very low bacterial count and made some very important observations with regard to the role of the different groups of organisms in the production of carbonyl compounds during bread fermentation. The following carbonyl compounds were analyzed — 2-hexanone, n-hexanal, isovalerylaldehyde, n-butyraldehyde, acetone, and acetaldehyde. A total of about 400 ppm of carbonyl compounds were found in both preparations, with very little difference in the qualitative and quantitative spectrum of the individual compounds. It was concluded that bacterial activity contributes little, if at all, to the formation of these compounds, which are probably derived from the activities of yeasts. Miller and associates (1961), who analyzed the carbonyl compounds in pre-ferments, found in addition to the previously mentioned compounds formaldehyde, isobutyraldehyde, methyl ethyl ketone, n-valeraldehyde, as well as 2-methyl-1-butanal. No attempt was made to trace the origin of these compounds. Other workers extended this list by adding propionaldehyde, n-hexaldehyde (Linko et al., 1962), acetoin, and diacetyl (Smith and Coffman, 1960). It seems, however, that the amount of carbonyl compounds detected in pre-ferments are much inferior to those found in conventional doughs.

The presence of esters in pre-ferments has been studied by a number of workers. Obviously, ethyl esters predominate. Ethyl formate, ethyl acetate, ethyl lactate, and 1,3-propanediol monoacetate were found in pre-ferments (Johnson et al., 1958; Smith and Coffman, 1960). However, there is little indication what part microbes take in the formation of these esters. Although many more compounds have been described in doughs, the absence of data regarding their occurrence in the ingredients makes it difficult to evaluate the part played by microorganisms. The role of bacteria in the formation of flavor in baked products could be better appreciated if chemical analysis could be performed on chemically leavened bread in comparison with bread prepared with pure yeast. The use of aseptic flour and sterile working conditions could be very helpful.

It should be pointed out that compounds produced by the different organisms during the panary fermentation contribute only in part to the final flavor of bread due to their great volatility and their disappearance during baking. Other flavor components are produced during crust formation to which microorganisms may contribute only indirectly, e.g. amino acids produced during the proteolysis of dough proteins will enhance the browning reactions that take place in the oven. Wiseblatt and Zoumutt (1963) studied aroma formation and the interaction of various amino acids with dihydroxyacetone, a product of fermentation, during baking (Table XXXVIII).

The formation of the crackerlike aroma by the action of proline on dihydroxyacetone seems to be of special importance. The compound responsible for the cracker aroma is probably 1,4,5,6-tetrahydro-2-acetopyridine:

which may be synthesized from proline and dihydroxyacetone (Hunter et al., 1969) or N-methyl-2-acetopyrrolidine:

which may be formed from proline and glycerol (Hunter et al., 1966).

A logical development in the microbiology of bread flavor would be the isolation and propagation of bread bacteria and the addition of these organisms to doughs. Such experiments have indeed been carried out by several groups (Miller and Johnson, 1958; Robinson et al., 1958). Cultures were isolated from various sources such as aged sponge dough, pre-ferments, and also from dairy material. Organoleptic evaluations of bread prepared with such starters were carried out by a taste panel and are summed up in Table XXXIX.

Certain bacteria did improve the taste and aroma of such breads, especially those prepared with lactic cultures. However, no analysis of flavor compounds of such loaves were recorded. Similarly, Carlin (1959) investigated the flavor fortification of straight and sponge doughs by the introduction of various bacteria from the genera of *Leuconostoc* and *Lactobacillus*, previously isolated from commercial yeast. A definite improvement of bread flavor has been described.

TABLE XXXVIII

AROMAS RESULTING FROM THE ACTION OF AMINO ACIDS
AND DIHYDROXYACETONE

Amino Acid	Resulting Aroma
Proline	Very strong, crackers, crust, toast
Lysine	Strong, dark corn syrup
Valine	Strong, yeast, protein hydrolysate
Alaline	Weak, caramel
Glutamic acid	Moderate, chicken broth
Aspartic acid	Very weak
Arginine	Very weak
Cysteine	Mercaptan, hydrogen sulfide
Hydroxyproline	Weak, like proline
Phenylalanine	Very strong, hyacinth
Leucine	Strong, cheesy, baked potato
Isoleucine	Moderate crust
Serine	Weak, vaguely like bread
Threonine	Very weak
Methionine	Baked potato
Glycine	Baked potato
Histidine	Very weak

SOURCE: After Wiseblatt and Zoumutt (1963).

TABLE XXXIX

AROMA DESCRIPTION OF FRESH BREADS PREPARED WITH
BACTERIAL STARTERS

Microorganisms Added	Odor Description of Fresh Bread
Standard loaf pre-ferments:	
Aerobacter aerogenes	Similar to standard
Bacillus coagulans	Stronger than standard
Bacillus laterosporus	Slightly stronger than standard
Lactobacillus plantarum	Strong, like fermentation
Micrococcus caseolyticus	Strong, like fermentation
Lactobacillus leichmannii	Mild, like fermentation
Lactobacillus casei	Mild, like fermentation
Aged sponge dough:	
Lactobacillus fermenti	Mild, like standard
Lactobacillus casei	Mild, like standard
Lactobacillus brevis	Cheeselike
Lactobacillus bulgaricus	Cheeselike
Lactobacillus plantarum	Cheeselike
Lactobacillus lactis	Strong, not acceptable

SOURCE: After Robinson et al. (1958).

The use of starter cultures of bacteria for the panary fermentation seems to have become a commercial fact in some of the
East European countries (Siver, 1975; Wlodarczyk et al., 1977).
Addition of liquid cultures to doughs (60 to 120% on flour
weight) shortened the production time by 60 to 70 minutes and
gave good results with regard to bread flavor, uniformity of
crumb porosity, and elasticity of rye and wheat rye breads.
Lactobacillus Sanfrancisco is currently manufactured by two
American firms for the production of sourdough French bread
and also for enhancing the flavor of white pan bread (Sugihara,
1977). The addition of fermented milk products such as yoghurt
in amounts of up to 20 percent of dough is claimed to improve
the flavor and nutritive value of baked products (Hill, 1974).
Also, a patent was granted (Sjostrom and Moriartry, 1971) describing the production of a "Yeast sensory stimulating factor"
by cultivation of a strain of *Lactobacillus casei* and the introduction
of the cell free broth (1 to 2%) into a dough, resulting in improved flavor of the baked products.

BREAD ENRICHMENT BY OTHER YEASTS. A number of workers
have suggested the improvement of the nutritional value of
breads with regard to their protein and vitamin content by the
incorporation of non-*Saccharomyces* types of yeasts. Kowalzuk
and Muszkatowa (1973) obtained good results by the incorporation of a yeast preparation made from *Candida (Torula)* species,
while Drobot and Stabnikova (1976) suggested the use of the
carotenoid-containing yeast *Rhodotorula glutinis* for a similar
purpose.

BREAD IMPROVEMENT BY MICROBIAL ENZYMES. In order to
somewhat lengthen the short shelf life of most breads and post-
pone as much as possible the staling process of baked products,
many authors attempted the use of amylolytic enzymes of microbial origin. The production of low molecular weight fragments from starch by α-amylases seems to retard the staling of
bread. Bacterial enzymes are probably best for this purpose due
to their thermal stability (Table XL). The *Bacillus subtilis* amylase
was found to be active even at temperatures of 90 to 95°C, which
occur in the inner portion of bread during baking (Miller et al.,
1953; Schultz and Uhlig, 1972; Velasco, 1975).

TABLE XL

THE EFFECT OF ENZYME ADDITION ON BREAD FIRMING

Enzyme source	Storage Time (hrs.)				
	18	42	66	90	114
Control	100*	140	180	220	225
Fungal	90	120	140	160	200
Bacterial	68	69	69	72	73

SOURCE: Adapted from Maga (1975).
* Firmness in grams.

Good results were also obtained by the use of an amylase preparation of the fungus *Aspergillus oryzae* (Kovacevic, 1974). The historical approach to the study of fermented products and their flavor evolution, i.e. microbial and chemical analysis, fermentation with pure cultures, and resynthesis of the process by means of a mixed culture, has been followed in the study of the flavor of baked products. Although the exact chemical composition of the flavor of such products is still uncertain, it appears that many of the flavor components are derived from bacterial fermentations, which promote the leavening activity of yeasts. The more recent attempts to introduce pure cultures of bacteria to the yeast mix may be regarded as a semisynthetic version of the original *Sauerteig* procedure.

BIBLIOGRAPHY

Bidmead, D. S., Grossman, J. D., and Kratz, P. D.: Roasted meat flavor and process for producing same. U.S. Pat. 3,394,016, 1968.

Carlin, G. T.: The fundamental chemistry of breadmaking. *Proc Amer Soc Bak Eng, 1:*56, 1958.

———: The fundamental chemistry of breadmaking. *Proc Amer Bak Eng, 2:*136, 1959.

Cole, E. W., Hale, W. S., and Pence, J. W.: The effect of processing variations on the alcohol, carbonyl, and organic acid contents of preferments for bread making. *Cereal Chem, 39:*114, 1962.

Davis, C. F. and Stephens, W. J.: What is fermentation loss? *Baker's Dig, 28:*5, 1954.

Drazhner, T. M., Raev, Z. V., Kovalenko, A. D., Kordyukova, N. S., and Komenko, E. V.: Improving maltase activity of baker's yeast manufactured at distilleries. *Tr Ukr Nauchno-Issledov Inst, 15:*46, 1973.

Drobot, V. I. and Stabnikova, E. V.: Improvement of the biological value of bread via enrichment with a biological preparation containing carotenes. *Izv Vys Uchebnyk Zav Pich Technol, 3:*57, 1976.

Galli, A. and Ottogalli, G.: Microflora of the sour dough of "Panettone." *Ann Microbiol Enzimol, 23:*39, 1973.

Heiman, W. and Drawert, F.: Zur Bildung pflanzlicher Aromastoffe. *Brot u Gebäck, 22:*133, 1968.

Hill, L. G.: Yoghurt-containing dough composition and baked product made therefrom. U.S. Pat. 3,846,561, 1974.

Höhn, E. and Solms, J.: Untersuchungen über Geschmacksstoffe der Hefe. *Lebensm Wissensch Technol, 8:*206, 1975.

Hunter, I. G., Walden, M. K., McFadden, W. H., and Pence, J. W.: Production of bread-like aromas from proline and glycerol. *Cereal Sci Today, 11:*493, 1966.

Hunter, I. G., Walden, M. K., Cherer, J. R., and Lundin, R. E.: Preparation and properties of 1,4,5,6-tetrahydro-2-acetopyridine, a cracker-odor constituent of bread aroma. *Cereal Chem, 46:*189, 1969.

Jackel, S. S.: Fermentation flavors of white bread. *Baker's Dig, 43:*24, 1969.

Johnson, J. A., Miller, B. S., and Curnutte, B.: Organic acids and esters produced in pre-ferments. *J Agric Food Chem, 6:*384, 1958.

Kohn, F. E., Wiseblatt, L., and Fosdick, L. S.: Some volatile carbonyl compounds arising during panary fermentation. *Cereal Chem, 38:*165, 1961.

Kovacevic, M.: Use of enzymes in breadmaking. *Hrana Ishrana, 15:*106, 1974.

Kowalzuk, M. and Muszkatowa, B.: The use of dried yeasts of the genus *Candida* and skim-milk for bread enrichment. *Zag Piek ZBPP, 18:*1, 1973.

Kurkela, R. and Uutela, P.: Aroma changes of thiamine-glutamic acid solutions during heating. *Lebensm Wissensch Technol, 5:*43, 1972.

Linko, Y., Miller, B. S., and Johnson, J. A.: Quantitative determination of certain carbonyl compounds in pre-ferments. *Cereal Chem, 39:*263, 1962.

Maga, J. A.: Bread flavor. *Crit Rev Food Technol, 5:*55, 1974.

————: Bread staling. *Crit Rev Food Technol, 5:*443, 1975.

Markova, J.: Qualitative and quantitative analysis of organic acids separated from rye flour, pre-ferment, dough and mix bread. *Sb Vyx Sk Chem Technol Pr Potr, 29:*123, 1970.

Matz, S. A.: *Bakery Technology and Engineering,* 2d ed. Westport, Connecticut, Avi, 1972.

Miller, B. S., Johnson, J. A., and Palmer, D. L.: A comparison of cereal, fungal, and bacterial alpha-amylases as supplements for breadmaking. *Food Technol, 7:*38, 1953.

Miller, B. S., Johnson, J. A., and Robinson, R. J.: Identification of carbonyl compounds produced in pre-ferments. *Cereal Chem, 38:*507, 1961.

Morrimoto, T.: Studies on free amino acids in sponges, doughs, and baked soda crackers and bread. *J Food Sci, 31:*736, 1966.

Otterbacher, T. J.: A review of some technical aspects of bread flavor. *Baker's Dig, 33:*36, 1959.

Pederson, C. S.: *Microbiology of Food Fermentations.* Westport, Connecticut, Avi, 1971.

Reed, G. and Peppler, H. J.: *Yeast Technology.* Westport, Connecticut, Avi, 1973.

Robinson, R. J., Lord, T. H., Johnson, J. A., and Miller, B. S.: The aerobic microbiological population of preferments and the use of selected bacteria for flavor production. *Cereal Chem, 35:*295, 1958.

Roiter, I. and Borovikova, L.: Level of aromatic volatile carbonyl compounds at various stages in the production of bread. *Izv Vyssh Uchebm Zaved Pishch Tekhnol, 5:*68, 1971.

Schulz, A. and Uhlig, H.: Einsatzmöglichkeiten von Alpha-amylasen in der Bäckerei unter besonderer Berücksichtigung der Bakterien-amylase. *Getreide Mehl u Brot, 26:*215, 1972.

Siver, V. E.: Preparation of rye flour doughs from M-1 liquid starter. *Khleb Kondit Promyshlenost, 12:*36, 1975.

Sjostrom, L. B. and Moriartry, J. H.: Yeast sensory stimulator and process of making same. U.S. Pat. 3,615,696, 1971.

Smith, D. E. and Coffman, J. R.: Separation and identification of the neutral components from bread pre-ferment liquid. *Anal Chem, 32:*1733, 1960.

Spicher, G. and Schöllhammer, K.: Vergleichende Untersuchungen über die in Reinzuchtsauer — und spontansauerteigen anzutreffenden Hefen. *Getreide Mehl u Brot, 31:*215, 1977.

Sugihara, T. F.: Non-traditional fermentations in the production of baked goods. *Baker's Dig, 51:*76, 1977.

Sugihara, T. F., Kline, L., and Miller, N. W.: Microorganisms of the San Francisco sour dough bread process. *Appl Microbiol, 21:*456, 1971.

Suomalainen, H., Dettwiller, J., and Sinda, E.: Alpha-glucosidase and leavening of baker's yeast. *Process Biochem, 7:*16, 1972.

Tanaka, Y. and Sato, T.: Fermentation of fructosides in wheat flour by baker's yeast. *J Ferment Technol, 47:*587, 1969.

Velasco, J. G.: Bacterial α-amylases and retention of bread flavor freshness. *Panadero Latinamer, 29:*4, 1975.

White, J.: *Yeast Technology.* London, Chapman and Hall, 1954.

Wiseblatt, L. and Zoumutt, H.: Isolation, origin and synthesis of a bread flavor constituent. *Cereal Chem, 40:*162, 1963.

Wlodaczyk, M., Zajdel, M., and Augustyniak, M.: Application of pure dairy cultures of yeasts and bacteria for leaven used in commercial bakeries. *Przegl Piek Cukier, 25:*26, 1977.

Chapter 7

FERMENTED BEVERAGES AND THEIR FLAVOR — I. WINE

Wine, though bitter, sweetens all bitterness.

—MOSES IBN EZRA *(Selected Poems)*

INTRODUCTION

D URING THE EARLY DAYS of civilization, wines were considered the number three beverage, after water and milk. Its early appearance in man's history may have been due to the fact that it was more the invention of nature than of man. Juices containing sugars start to ferment spontaneously; sometimes even mature fruits release the pleasant winy fragrance that accompanies the alcoholic fermentation. The intoxication and erratic behavior of animals after consumption of such fruit is well known.

Alcoholic fermentations are primarily initiated by the heterogeneous yeast population that appears in fruit juices, of which the must obtained from crushed grapes is the most important. Such musts carry a heavy load of yeasts (about 10^4 to 10^6 ml), represented by many genera and species. An example of this heterogeneous population may be seen in Table XLI.

The classification of yeasts is very difficult and should be left to experts. Information may be retrieved from the comprehensive key prepared by Lodder (1970). Here it is sufficient to point out that in addition to the morphological features (size, shape, mode of budding, mycelium formation), yeasts are classified according to their fermenting capacity and utilization of sugars for assimilation and fermentation. The appearance of ascospores (the sexual spores of the Ascomycetes, see Chapter 1) is also an important feature. An exhaustive survey of yeasts that were isolated from musts can be found in the review by Kunkee and Amerine (1970).

Although the yeast population may vary according to climate

173

Flavor Microbiology

TABLE XLI
YEASTS IN FRESH GRAPE JUICE

Type	% of Total
Apiculate yeasts	58-76
Torulopsis	13-19
Saccharomyces	0-12
Film yeasts	3-10
Rhodotorula and other	1-4

Source: After Goto and Yokotsuka (1977).

and topographic conditions, it is striking that most of these groups have been described in such diverse geographic areas as in the Mediterranean countries, central Europe, California, and Japan. Practically in all cases, the population is predominantly that of the poorly to medium fermenting yeasts, such as members of the apiculate group *(Kloeckera, Hanseniaspora)* and the more oxidative film yeasts *(Hansenula, Candida, Pichia)*. Although members of the highly fermentative group of yeasts *Saccharomyces (S. cerevisiae, S. oviformis)* are usually present in fresh musts, their number is relatively small. This situation changes rapidly as a result of the fermentation that sets in and the difference in ethanol resistance of the various yeast types (Table XLII). The slower fermenters or less ethanol resistant yeasts yield to the high fermenters or more alcohol resistant organisms so that by the end of a normal fermentation the number of apiculate yeasts and other poor fermenters is reduced to a minimum, while the vigorous alcohol producers dominate (Domerq, 1957). This, however, is not always the case. Very often and because of a variety of reasons (small numbers of *Saccharomyces* yeasts, unfavorable environmental conditions, etc.) the desired yeast microflora does not become predominant, as a result of which, the freshly fermented must does not reach the suitable alcohol concentration ("stuck wine"). Such wines very often undergo spoilage due to yeast autolysis or proliferation of undesirable bacteria, both affecting the flavor of the product.

The use of sulfur dioxide as a protective agent against the proliferation of undesirable yeasts and bacteria has been known

TABLE XLII
MAXIMUM ETHANOL PRODUCTION BY VARIOUS YEASTS

	% vol
Candida pulcherrima	2.0
Kloeckera apiculata	3.5-5.0
Kloeckera africana	10.2
Saccaromyces pastorianus	7.9-8.4
Saccharomyces rosei	5.3-8.5
Torulopsis bacillaris	8.1-8.9
Brettanomyces bruxellensis	9.1
Saccharomycodes ludwigii	10.7-12.0
Schizosaccharomyces pombe	12.2
Saccharomyces oviformis	11.4-12.8
Saccharomyces cerevisiae, Bakers' yeast	15.1
Saccharomyces cerevisiae, var. *ellipsoideus*	14.5-16.5
Saccharomyces cerevisiae, Tokay strain	12.8
Saccharomyces cerevisiae, Burgundy strain	18.4

Source: Compiled from Malan (1956), Ferreira (1959), and Rosini (1975).

for many generations (the "burning of sulfur"), many of the so-called wine yeasts having developed a higher resistance to such treatments. Nevertheless, even these precautions are not always sufficient. In many cases, especially in countries with warm and hot climates, wine makers prefer the addition of a massive starter made of a selected strain of a *Saccharomyces* in addition to the sulfur treatment. Although evidence is available that wines made with added selected yeasts are qualitatively superior to wines made by natural fermentation (Rankine and Lloyd, 1960), many established wine cellars, especially in areas with moderate climates, still prefer the natural process (Radler, 1973), thus avoiding the production of what may be named "monotonous" wines, those without the ups and downs of a natural fermentation.

The composition of the natural yeast flora of grapes may be affected by the nature of the fungicides employed in the vineyards. Thus, Sapis-Domerq et al. (1976) observed a decrease in the number of *Saccharomyces* species in places sprayed with the fungicide Euparene.[®]

WINE FLAVOR

Before going into the specific contribution of microorganisms to wine flavor, it is appropriate to dwell first on the general characteristics of the taste and aroma of wine. Like other fermented food products, the flavor of wine is a complex one, consisting of a large number of compounds derived from a wide variety of sources, viz. raw materials: grape varieties and degree of maturation; agrotechnical and climatic considerations — soil, water, sun that prevailed during grape cultivation (year of vintage!); techniques employed for must production, harvest, and transport; the type of fermentation (containers, color extraction from skin, extent of sulfitation); termination of the fermentation process; fortification; amelioration; secondary fermentations; and finally the type of aging (bottles, wooden containers) and storage conditions. All of these factors will ultimately affect the flavor of the product. We shall not go into the technological aspects of wine production for which excellent textbooks in several languages are now available (Ribereau-Gayon and Peynaud, 1961, 1964; Troost, 1961; Amerine et al., 1967) but rather limit our discussion to the microbiological aspects during the different phases of wine fermentation with emphasis on the formation of the characteristic flavor or, as wine lovers prefer to call it, the typical "bouquet" of a specific wine.

Wines occupy a special position in the field of flavor chemistry as there is scarcely another product that is valued, appreciated, or criticized for its flavor qualities as much as wine. The high prices that are paid for an outstanding bottle of wine may be considered ample evidence for the outstanding importance of its flavor qualities. In fact, the long history of wine making and the detailed customs or rites accompanying the consumption of wine in various cultures have led to a wealth of expressions for the description of wine flavor, which are hardly met with in other beverages: sweet and dry, smooth and astringent, light and heavy, viscous, acid, fruity, flowery, interesting, complex, and many other expressions used by the connoisseur — all in harmony with each other to create the subtle nature of a specific bottle. Since the appreciation of a mixture of flavor components, many of which appear only in trace amounts, is more the result of a personal description than a matter of qualitative or quantitative analysis, the quality of wine and its flavor evaluation are estimated and characterized by an expert enologist by tasting and not by gas chromatography. Chemical analysis, up to now and in the foreseeable future, can serve only for the accumulation of facts in support of the conclusions reached by taste experts.

A sparkling liquidlike wine might be considered an excellent subject for the analysis of flavor. However, the high alcohol, watery product makes the isolation of minor compounds a very difficult task for the flavor chemist. These difficulties have been described in detail by Webb (1967). In spite of these technical difficulties, over 100 compounds that may contribute to the flavor of wine have been described. In this chapter, we shall limit our interest to those compounds that are considered metabolic products of microorganisms and are directly or indirectly involved in the formation of taste and bouquet of wine. Contrary to the panary fermentation, in wine making the product of major importance is ethanol, which is the key determinant of wine character. The amount of alcohol produced during the fermentation depends first of all on the sugar concentration available. Grapes are probably the sweetest of all fruits and in suitable climates may reach values above 20° Brix. In some areas where the grapes are left to dry before crushing, much higher sugar concentrations may be obtained (30° Brix and over). Chemists describe the flavor characteristics of ethyl alcohol as having "a pleasant odor and sharp taste" (Windholz, 1976). Considering the high alcohol content of wines, the highest of any undistilled beverage, ethanol constitutes the backbone of wine flavor.

Another much less known attribute of ethanol in wine flavor is its effect on the perception of other bouquet-forming principles. According to Rankine (1967), the threshold value of a large number of higher alcohols, an important flavor component of alcoholic beverages (see following text), is increased over a hundredfold as compared to the values obtained in distilled water for the corresponding alcohol.

The production of ethanol by fermentation of yeasts depends also on the nature of the carbohydrate substrate. While most yeasts ferment glucose at a higher rate than fructose (although mature grape must usually contains more fructose than glucose), some wine yeasts, especially the well known Sauterne strains, ferment fructose at a higher rate (Sols, 1956).

Sugar fermentations yield not only ethanol and CO_2 but also a variety of other metabolic products, which may be related to the formation of wine flavor and bouquet even if present in only minute concentrations. Quantitatively, the most important by-products of such fermentations are glycerol and succinic acid (Amerine, 1954). According to Oura (1971), about 5 g of glycerol and 0.7 g of succinic acid are produced for every 100 g of ethanol. In wine containing about 10 percent (weight) ethanol, one would expect about 0.5 percent glycerol. In a compilation of values of glycerol in wines from forty-five different sources, Amerine (1954) reports average concentrations from 0.32 percent to 1.65 percent. The higher values probably resulted in wines prepared from moldy grapes. The ratio of alcohol/glycerol in wines ranged from 8 to 15. According to Ribereau-Gayon and Peynaud (1964), *Zygosaccharomyces acidifaciens* is outstanding in its glycerol-producing capacity, yielding up to 1.5 percent during an alcoholic fermentation. Environmental factors probably affect glycerol formation. Lower temperatures and high sulfitation lead to higher values. Drzaga et al. (1976) obtained 12 g/liter glycerol in musts fermented in the presence of 150 mg SO_2/liter. Nonsugar carbon sources may also contribute to the formation of glycerol. According to Liebert (1971), both malate and tartarate may be utilized for glycerol systems provided sufficient amounts of thiamine are available. Glycerol imparts smoothness to wine, ameliorates the sharp, burning taste of ethanol, and may therefore be considered as an important factor in the buildup of wine flavor. Its sweet taste and viscous character undoubtedly affect the quality of wine, especially what the enologists call the "body" of the beverage.

Succinic acid formed during the alcoholic fermentation will contribute to the acidity of wine. Levels up to 0.15 percent have been recorded (Amerine and Cruess, 1967). Earlier reports on the relationship between the secondary products of fermentation as expressed in the Genevois equation (1960):

$$G = 5s + 2a + h$$

where G = moles of glycerol, s = succinic acid, a = acetic acid, and h = acetaldehyde were later modified to include other fermentation products also (acetoin, butylene glycol) which have little to do with wine flavor.

Acetaldehyde, a normal by-product of the fermentation, may accumulate in wine as a result of ethanol oxidation or the activity of film yeasts. At a concentration of 0.5 g/liter, wines are no longer marketable. Acetaldehyde concentrations will usually be in the range of 20 to 200 ppm, exceeding these figures in wines with an oxidative phase, like sherry. According to Hinreiner et al. (1955), the flavor threshold level of acetaldehyde in red and white table wines is 100 and 125 ppm respectively.

Acetic acid is a very important factor in wine. It constitutes the major volatile acid in this fermentation and may not exceed 0.12 to 0.14 g per 100 ml in order to be marketable. Although acetic acid is the major product of bacterial activity by many species of *Acetobacter*, it may also be formed directly by the oxidation of acetaldehyde as well as by the activity of certain yeasts. According to Cappucci (1948), apiculate yeasts, as well as some of the *Saccharomyces* yeasts *(Zygosaccharomyces acidifaciens)*, may produce considerable amounts of acetic acid which may reach up to 0.369 percent. It seems that the levels of acetic acid in normal fermentation change according to the redox-potential of the must. During the initial stages, levels of acetic acid are considerably higher than toward the final stages (Ribereau-Gayon and Peynaud, 1946).

The acetification of alcohol to acetic acid carried out by a variety of *Acetobacter* species is a typical spoilage process known from the early days of history. This, however, has resulted in the production of vinegar, an essential flavor ingredient of many food preparations. Many different raw materials originating in alcoholic fermentations or distilled products may be employed for this process: wine, malt, cider, spirits, all having in common the alcohol produced by a yeast fermentation. The stoichiometry of alcohol oxidation by *Acetobacter* is mainly according to

$$CH_3CH_2OH \xrightarrow{O_2} CH_3COOH + H_2O$$

clearly indicating an intense oxidative process which, considering the very high ethanol content used (10 to 12% w/w), requires suitable oxygen supply. The technology of vinegar production has been amply described in textbooks, e.g. Greenshields (1978). As to flavor, vinegar contains many products and by-products of the alcoholic fermentation: glycerol, ethanol, and many aroma products derived from the plant raw materials not yet fully assessed. Many of these will not be found in synthetic vinegar. The high volatile acetic acid content of commercial vinegar (5 to 7%) masks many flavor components, no doubt; nevertheless, most people can distinguish between different products according to their alcoholic base.

The titrable acidity of wines, ranging between 0.40 and 0.70 g/100 ml (as tartaric acid), comprises a large variety of acids, only a fraction of which is derived from microbial activity through the glycolytic or citric acid cycles, the major part stemming from the must. We shall return later to these acids as a substrate for secondary fermentations.

From what has been said so far, it looks as if the desirable components produced by yeasts during the vinification process are chiefly ethanol, hence, the flavor profile of wine is determined mainly by the quality (variety, year of vintage) of the grapes and less by microbial activity. This view was held by many workers but may be changing as a result of new information on the nature of wine flavor and bouquet which has become available after the introduction of more sophisticated methods of analysis.

The number of chemical compounds isolated and identified in wines has increased steadily during the last two decades. Drawert and Rapp (1969) described over 180 compounds in musts and wines. Webb and Muller (1972) enumerate over 300 compounds in alcoholic beverages. Even these reviews did not exhaust the subject, as can be seen from the new compounds described more recently (Bertuccioli and Viani, 1976; Schreier et al., 1977). In spite of the enormous advances in analytical procedures, these techniques have so far not been sufficiently employed for the analysis of flavor components produced by pure cultures in grape juices of different varieties. A cooperative effort including microbiologists and analytical chemists would yield much information on this subject. We cannot always distinguish between contribution of grape juice to wine flavor and those compounds transformed or synthesized by the yeast flora.

Microbiologists have considered for a long time the spontaneous fermenting grape must as fertile ground for the isolation of a large variety of yeasts. Following Koch's principles for the relationship between a pure culture and its chemical activities, much information on the contribution of various yeasts to minor flavor components should become available (Park and Bertrand, 1974). In the rest of this chapter, we shall limit our discussion to those compounds where evidence is already available regarding their production via microbial activity.

Among the keto acids, pyruvic acid and α-keto glutaric acid seem to be quantitatively the most important acids that accumulate during vinification. Lafon-Lafourcade and Peynaud (1966) found up to 785 mg/liter of pyruvic acid and up to 146 mg/liter of α-keto glutaric acid in wines of different vintages. Yeasts related to the vinification process seem to have a different ability to form these keto acids. Low values were obtained with wild yeasts such as *Kloeckera apiculata* or *Hansenula anomala* (less than 100 and 30 ppm respectively), while strong alcohol-producing yeasts formed higher levels; in the case of *Schizosaccharomyces pombe*, up to 830 and 284 ppm respectively. Similar results were obtained by Rankine (1967). The accumulation of pyruvic acid during the alcoholic fermentation varied considerably with the strain employed and the type of must fermented. With twelve different wine yeasts, a range of 8 to 120 ppm (average 40 ppm) was observed; highest values were obtained with *S. oviformis* No. 723. There seems to be less decarboxylation with fermentations run at higher pH values, resulting in a larger accumulation of pyruvate. In an artificial juice fermented with *S. oviformis,* maximum values were obtained after 6 days (146 ppm at pH 3.3) and after 15 days (513 ppm at pH 4.8). Also, the formation of α-keto glutaric acid seems to vary with the yeast strain and is largely affected by environmental conditions. An increase in the fermentation temperature increased the levels of this acid (Rankine, 1968). We know little about the importance of these acids in wine flavor, but they may considerably affect the stability of wines due to their reaction with SO_2. It is recommended to run wine fermentations with yeasts selected for their low production of keto acids at low temperature and pH values.

Another group of compounds that appears during vinification and for which ample evidence for the involvement of microbial activity is available is the group of esters. The importance of such esters to the formation of wine bouquet has already been studied for many years. Ethyl acetate is the most prevalent ester in fermented must and, as pointed out by Amerine (1954), constitutes the only ester important from the standpoint of taste and aroma. Other esters such as ethyl caproate, heptanoate, caprylate, decanoate, pelargonate, laurate, isoamyl acetate, and others have been identified in various wines but may be of greater importance in distilled beverages (Crowell and Guymon, 1969).

The content of ethyl acetate in wine varies according to the grape variety and the vinification process. Ribéreau-Gayon and Peynaud (1961), who analyzed the composition of wines during several decades, found wines to contain generally between 44 and 176 mg/liter ethyl acetate; wines seem to tolerate up to 200 mg/liter of ethyl acetate, above this value a spoiled character appears (Amerine and Cruess, 1967). It is now generally agreed upon that the formation of these esters is due to the metabolic activities of microorganisms and little, if at all, to chemical esterification. Ribéreau-Gayon and Peynaud (1961) distinguish between five different groups of microorganisms with regard to their production of ethyl acetate: (1) the true wine yeasts of the genus *Saccharomyces,* which never produce more than 50 mg/liter; (2) yeasts from the group of *Candida pulcherrima, Kloeckera africana,* and *Brettanomyces* spp. which may accumulate up to 100 mg/liter; (3) *Saccharomycodes ludwigii,* which differs from the other yeasts in the fact that its esterogenic activity is greatly enhanced under anaerobic conditions (grape juice fermented with *Saccharomycodes ludwigii* has a very pronounced odor of this ester — up to 200 mg/liter; (4) the widely occurring apiculate yeasts of the group *Kloeckera-Hanseniaspora,* which produces over 300 mg/liter of the ester; and (5) the group that comprises the yeasts such as *Pichia* spp. and more importantly, *Hansenula anomala,* which apparently has the strongest esterogenic activity. The latter was found to produce up to 900 mg/liter of ethyl acetate and thus may be detrimental to the process of vinification. With the exception of *Saccharomycodes ludwigii,* all esterogenic activity is promoted by contact with air. Under these conditions, various species of *Acetobacter* may proliferate with a concomitant increase in the formation of ethyl acetate. Tabachnick and Joslyn (1953) have studied the kinetics of ethyl acetate formation and found that levels of the ester decrease with time under aerobic conditions. This seems to be in accordance with the observation of Wahab et al. (1949) who found that on aging, the ester content of wines decreases substantially with a constant improvement of wine bouquet.

Another aspect of yeast microbiology has drawn the attention of many enologists. Yeasts differ in their capability of producing foam during the active fermentation of musts. Since this is not a desirable feature of wine yeasts, attempts have been made to select strains with reduced foaming ability (Thornton, 1978). Nonfoaming mutants of *S. cerevisiae* are now available. According to Dittrich and Wenzel (1976), this character may be due to the reduced ability of such yeasts to produce hydrophobic proteins on the surface of the cell wall. It would be interesting to study the flavor profile of wines fermented with nonfoaming yeast starters.

FUSEL OIL ALCOHOLS

Among the components that occupy an important position in wines and which are definitely the result of microbial activity, we have to mention the so-called "fusel oil" alcohols. These alcohols (excluding ethanol, glycerol, as well as methanol, the latter not being a fermentation product) include a number of higher aliphatic alcohols, which contribute much to the "hangover" sensation after consumption of alcoholic beverages. These usually comprise the following four alcohols:

	Boiling point, °C
n-propanol	97
isobutanol (2-methyl-1-propanol)	108
active amyl alcohol (active pentanol, 2-methyl-1-butanol)	128
isoamyl alcohol (isopentanol, 3-methyl-1-butanol)	132

Other aliphatic alcohols like n-butanol have also been described. The most objectionable with regard to taste and odor is that of isoamyl alcohol, having a pungent and repulsive taste. The average fusel oil content of wines has been examined by a large number of workers. In general, the content of these higher alcohols is greater in red wines than in white wines. Guymon and Heitz (1952) give a mean value of 250 ppm for white Californian table wines versus 287 ppm for red table wines, while Peynaud and Guimberteau (1962) find 309 ppm and 394 ppm (mean values) for the corresponding types of a large number of French wines. Although it is generally agreed that the fermenting yeast is responsible for the formation of higher alcohols (Guymon, 1966), until the 1950s little was done to examine the fusel-oil-producing capacity of various yeast strains and the environmental conditions that affect the production of these alcohols during vinification.

Webb and Ingraham (1963), in their exhaustive review of the problem of fusel oil, summarize earlier observations by stating that although various yeast genera comprising a number of wild yeasts produce different amounts of fusel oil (Table XLIII), little variation could be found among the naturally occurring wine yeasts.

Nevertheless, Webb and Kepner noted considerable differences in the composition of fusel oil of wines that were fermented with five different strains of wine yeasts under otherwise identical conditions. These differences are striking because of the analytical procedure (GLC) that enabled the differential determination of each of the fusel oil components (Table XLIV).

TABLE XLIII

PRODUCTION OF FUSEL OIL BY YEASTS OF DIFFERENT FERMENTATION
CHARACTER UNDER NONAERATED CONDITIONS

Yeast	Ethanol Formed % vol.	Fusel Oil Formed mg/100 ml
Pichia membranefaciens	0.1	0.3
Hansenula anomala	4.3	4.3
Candida albicans	2.9	3.8
Kloeckera magna	4.2	3.2
Debaryomyces kloeckeri	0.2	1.1
Saccharomyces beticus	12.3	4.7

SOURCE: After Guymon et al. (1961).
NOTE: Aeration apparently increases the production of fusel oil.

TABLE XLIV

FUSEL OIL FRACTIONS IN WHITE WINE (SAUVIGNON BLANC) FERMENTED
WITH DIFFERENT WINE YEASTS (% OF TOTAL FUSEL OIL)

Yeast	n-propanol	Isobutanol	Active Amyl Alcohol	Isoamyl Alcohol
Burgundy strain	18.2	12.4	12.0	57.4
Montrachet strain	2.6	2.7	16.5	78.2
Jerez (*S. beticus*)	20.2	8.4	4.5	66.9
Muscat Raisin strain	0.7	4.9	19.3	75.1
Zinfandel strain	2.3	15.6	16.8	65.3

Source: After Webb and Kepner (1961).

Variations in the capacity to produce fusel oil components by various strains of good ethanol producers were also observed by Rankine (1967).

The following ranges were observed—

n-propyl alcohol:	minimum	13 ppm (*S. carlsbergenis* No. 731)
	maximum	106 ppm (*S. cerevisiae* No. 350)
isobutyl alcohol:	minimum	9 ppm (*S. fructuum* No. 138)
	maximum	34 ppm (*S. chevalieri* No. 317)
amyl alcohols:	minimum	115 ppm (*S. cerevisiae* No. 213)
	maximum	262 ppm (*S. cerevisiae* No. 727)

The production by the various yeasts relative to one another seems to be consistent so that it would be justified to designate certain yeasts as high or low producers of some of the higher alcohols. Furthermore, it may be concluded that different strains of the same species seem to differ substantially in their ability to produce these alcohols. Guymon (1966) has also shown that mutants of *S. cerevisiae* would produce different amounts of the various higher alcohols according to their blocks in the biosynthetic pathway of the corresponding alcohol. The effect of fermentation temperature on the formation of fusel alcohols was studied by Ough and coworkers (1966). The synthesis of fusel alcohols may be indirectly affected by the use of fungicides, probably via their effect on the composition of the fermentation flora. Sapis-Domercq et al. (1976) reported an increase in the content of fusel oil alcohols from 343 mg/liter (control) to 631 mg/liter of wines derived from grapes that had been sprayed with the fungicide F83675.

Since, as has been pointed out earlier, ethanol dramatically increases the flavor threshold of these alcohols, the positive effect of yeast strains (that produce reasonable amounts of fusel oil) on the formation of a typical wine flavor must not be overlooked. Indeed, Webb (1967b) stated that "the group of alcohols is usually present at concentrations low enough so that the sensory impression is not unfavorable." In fact, some wines may be characterized by high concentrations of fusel alcohols. Harvalia et al. (1976), who examined the fusel oil composition of a number of Greek wines, report a large variation in the concentration of amyl alcohols which exceeded 300 mg/liter in certain types of Greek wines (Rhoditis, Monemvassia, Xynomavron). Ramos and Gomes (1974) found 650 to 800 mg/liter of higher alcohols to be typical for genuine port wines.

Biosynthesis

The metabolic pathways that lead to the formation of fusel oil alcohols have been the subject of research of many investigators. In the early days it was believed that these alcohols had little to do with the alcoholic fermentation itself but were rather the byproducts of bacterial contaminations. This view was later abandoned. Ehrlich (1907), Neubauer and Fromherz (1911), as well as other workers, suggested that the formation of these alcohols is connected with the metabolism of certain amino acids and is carried out by a series of reactions in which an amino acid becomes deaminated oxidatively, decarboxylated, and reduced to the corresponding alcohol:

$$
\begin{array}{ccccccc}
R & & R & & R & & R \\
| & -NH_3 & | & -CO_2 & | & NADH_2 & | \\
CH \bullet NH_2 & \longrightarrow & C=O & \longrightarrow & CHO & \longrightarrow & CH_2OH \\
| & +1/2\ O_2 & | & & & & \\
COOH & & COOH & & & &
\end{array}
$$

In accordance with this scheme,

valine	$\rightarrow$	isobutanol
isoleucine	$\rightarrow$	active amyl alcohol
leucine	$\rightarrow$	isoamyl alcohol

Transamination reactions with keto acids may take place in the first step (SentheShanmugathan, 1960). Since the amino acid

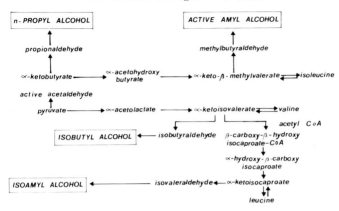

Figure 24. Scheme for the Biosynthesis of Fusel Oil Alcohols. After Webb and Ingraham (1963).

content of grape juice is not sufficient for the production of these alcohols at concentrations occurring in wines, it was postulated that other sources must also be available. Further, n-propanol does not correspond to any natural amino acid. As a result of studies with tracer techniques and mutants of the various pathways, it seems certain that many of the fusel oil alcohols are derived from precursors during the biosynthesis of these amino acids as carried out during growth and assimilation of the fermenting yeasts (Webb and Ingraham, 1963; Chen, 1977). A scheme for the biosynthesis of fusel oil alcohols is given in Figure 24.

Goranow (1976) studied the kinetics of fusel oil formation by GLC studies of samples taken daily from fermenting juice of Bulgarian musts. While the isoamyl alcohols as well as n-propanol could be detected virtually as soon as the fermentation started, isobutanol appeared only on the fourth day of the fermentation when about 30 percent of the sugars were consumed. A somewhat different type of kinetics was described in Italian wines (Fig. 25).

A correlation between the alcohol dehydrogenase activity of cell free extracts of wine yeasts and their potential for the production of the main fusel oil alcohols was established (Singh and Kunkee, 1976).

Another alcohol that appears in small concentrations in wines

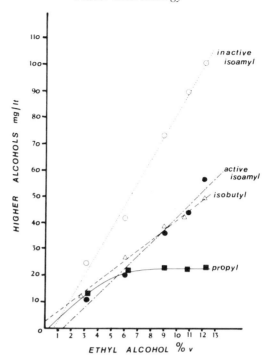

Figure 25. Formation of Higher Alcohols During Wine Fermentation. After Usseglio-Tomasset (1971).

is phenylethyl alcohol (β-phenethanol). It is of particular interest because it is probably related to the flowery character of the wine.

This alcohol with its rosy odor appears in wines at concentrations of 10 to 75 ppm (Sapis and Ribereau-Gayon, 1969) and may be perceived by many taste individuals at these levels. According to Rankine and Pocock (1969), yeasts differ in their biosynthetic capacity for the production of this alcohol (Table XLV).

Since the same yeast strain yields different amounts of phenylethanol in different varieties of grape juice, it may be assumed that this is due to different amounts of a precursor (phenylpyruvic acid?) present in the juice. N-hexanol, which

TABLE XLV

FORMATION OF β-PHENYLETHANOL AND N-HEXANOL BY WINE YEASTS IN
FILTER STERILIZED GRAPE JUICE

Yeast	β-phenyl ethanol* ppm	n-hexanol* ppm
S. *fructuum,* 138	7.8	3.1
S. *cerevisiae,* 161	5.6	3.3
S. *cerevisiae,* 275	20.0	2.9
S. *cerevisiae,* 348	7.6	3.1
S. *cerevisiae,* 350	8.9	2.9
S. sp., 719	7.7	3.4
S. *oviformis,* 723	8.9	3.6
S. *cerevisiae,* 729	9.5	3.4

SOURCE: Adapted from Rankine and Pocock (1969).
* Mean results of 3 grape varieties.

appears in much smaller amounts, is less affected by the yeast employed, although longer contact times with grape skin increases its concentration significantly. N-hexanol is probably produced from hexene-2-al-1, derived from linolenic acid through the reductive activity of yeasts (Drawert et al., 1965). It is claimed to impart a somewhat woody or foreign character to wines. While phenylethanol may improve the quality of wines, tasters seem to differ in their ability to discern the difference between various levels of this alcohol. The threshold values given were between 30 and over 200 ppm.

The addition of amino acids to fermenting musts was studied by Cerutti et al. (1978). Both phenylalanine and tryptophan (500 mg/liter Nosiola musts) improved the bouquet of such wines, probably because of higher levels of phenylethanol. An increased phenylethanol content was described in wines spontaneously fermented when compared to the same must fermented with pure culture wine yeasts. Differences of up to eight times in phenylethanol concentrations were observed in a number of comparative experiments. There is no doubt that this constitutes an important advantage of the traditional wine fermentation procedure which many wine makers refuse to abandon (Sponholz and Dittrich, 1974). Small amounts of 2-butanol have been detected in some wines (4 to 5 mg/100 ml ethanol). Since this alcohol has a somewhat oily-vinous, sweetish odor, it may contribute to the character of wines. Much higher concentrations (over 200 mg/100 ml ethanol) were found in certain brandies (especially in those made of wine remnants) where yeasts and lactic acid bacteria *(Lb. brevis)* could thrive. 2-butanol is probably produced by the reduction of 2,3 butanediol (Hieke and Vollbrecht, 1974).

THE MALO-LACTIC FERMENTATION

The nonvolatile acids of musts comprise malic acid and tartaric acids in roughly equal proportions. Much higher ratios of tartaric to malic acids (up to 4.04) were recorded in some Arizona grape varieties (Philip and Nelson, 1973). Titrable acidities of musts vary according to the ripeness of the grapes and may reach up to 0.8 percent, expressed as tartaric acid. Both acids undergo changes during the vinification process as a result of the activities of both yeasts and bacteria. Among the nonvolatile organic acids formed during fermentation, lactic acid occupies a very important position, both from the practical standpoint and biochemical interest. Lactic acid occurs in all wines that have undergone vinification. Wines that have had no bacterial contamination usually contain only very low levels (up to 0.06% of lactic acid), predominantly that of the D(−) isomer (Peynaud et al., 1966). The formation of lactic acid by various yeasts has been studied under controlled conditions of vinification (Peynaud et al., 1967). Considerable variations could be observed among the species of *Saccharomyces. S. rosei* gave the lowest level, 101 to 135 mg/liter, while most of the other saccharomycetes produced around 200 to 400 mg/liter. An interesting exception was that of *S. veronae*, which produced up to 1800 mg/liter lactic acid — most of which was that of the L(+) isomer.

The most significant feature of lactic acid in wines, with all its organoleptic implications, is not due to the activity of yeasts, however, but to the formation of lactic acid by certain groups of bacteria. Since the early days of wine microbiology, it was known that certain bacterial processes led to a change in the acidity of wines after the alcoholic fermentation was completed. A reduction of acidity is of great importance in many cases of table wines that are produced from juices particularly rich in total acidity.

L-malic acid is the major acid that undergoes reduction. This relatively strong dicarboxylic acid is readily attacked by many lactobacilli, which thus reduces the acidity of the wine by the formation of the relatively weak monocarboxylic lactic acid according to

$$
\begin{array}{c}
\text{COOH} \\
| \\
\text{OH---CH} \\
| \\
\text{CH}_2 \\
| \\
\text{COOH}
\end{array}
\quad
\begin{array}{c}
\text{Mn++} \\
\text{Malic dehydrogenase} \\
\text{(decarboxylating)} \\
\xrightarrow{\hspace{1.5cm}} \\
\text{NAD} \quad \text{NADH}_2
\end{array}
\quad
\begin{array}{c}
\text{COOH} \\
| \\
\text{C=O} \\
| \\
\text{CH}_3
\end{array}
\quad
\begin{array}{c}
\text{Lactic} \\
\text{dehydrogenase} \\
\xrightarrow{\hspace{1.5cm}} \\
\text{NADH}_2 \quad \text{NAD}
\end{array}
\quad
\begin{array}{c}
\text{COOH} \\
| \\
\text{HO---CH} \quad + \quad \text{CO}_2 \\
| \\
\text{CH}_3
\end{array}
$$

L(-) MALIC ACID   PYRUVIC ACID   L(+) LACTIC ACID

This decrease in malic acid and increase of lactic acid as a result of microbial activity is known as "malo-lactic fermentation." This fermentation usually develops during the last stages of the alcoholic fermentation or after the exhaustion of the fermentable sugars. It is more common with red table wines than with white wines. This is probably due to the higher acidities usually found in white wines and to the higher rates of sulfitation employed. On the other hand, red wines, which have been fermented for some time on the skins for color extraction, are usually richer in materials that promote the development of these organisms. This also explains the comparatively late appearance of the malo-lactic bacteria, which develop after yeast autolysis has taken place, providing many essential nutritive substances required by these bacteria (Lüthi and Vetsch, 1959). The malo-lactic fermentation does not always take place. In view of the high concentrations of ethanol, acidity, and the presence of SO₂, which constitute a rather hostile environment for such bacteria, this is not surprising.

The ability to ferment malic acid with the production of lactic acid and carbon dioxide found with many lactic organisms cuts across many generic lines. Although strains of *Lactobacillus*, *Pediococcus*, and *Leuconostoc* have been isolated from various malo-lactic fermentations, the *Leuconostoc* spp. are probably the most important organisms of this fermentation. Rankine (1977) found these organisms in almost all of the Australian red table wines examined. *Lc. oenos* (Garvie, 1967) is the most outstanding organism, being both resistant to high ethanol concentrations and fairly low pH values. This organism has received a species status in the last edition of *Bergey's Manual* (1974). Because of the erratic nature of the malo-lactic fermentation, attempts have been made to promote this fermentation by inoculation with a selected strain of *Lc. oenos*, e.g. the Californian ML34 (Kunkee, 1974). Freeze-dried preparations of a *Lactobacillus* have also become available (Equilait, France); so far, however, little information is available about the feasibility of such inoculations on a commercial scale. Another approach would be the use of malic acid degrading yeasts. Benda and Schmitt (1966) have suggested the use of the fission yeast *Schizosaccharomyces pombe* for the alcoholic fermentation of high acid wines since this organism also efficiently ferments malic acid with the production of ethanol and CO_2 even at very low pH values (Yang, 1973; Huglin et al., 1976). Other organisms may also degrade malic acid, e.g. *Kl. javanica* (Goto et al., 1978). Also *S. bailii* is an efficient scavenger of malic acid but cannot be recommended for use because it produces various off-flavors in wine (Radler et al., 1971).

Before going into some further details about the biochemical changes that take place during the malo-lactic fermentation and its effect on the flavor qualities of wines after such a secondary fermentation, a few words are appropriate about the biophysical nature of the reaction. There has been some controversy with regard to the thermodynamic likelihood of this reaction. Since all the NADH formed during the decarboxylation reaction leading to pyruvic acid is thereafter reoxidized to NAD for the reduction to lactic acid, Schanderl (1943) believed that the reduction of malic acid is an energy-consuming process. Kunkee (1967), however, calculated the changes in free energy, assuming the formation of carbonic acid (and not CO_2) as an end product and concluded that basically this is an exergonic reaction (-6Kcal/mole), with an equilibrium at the side of the end products. However, this energy does not seem to be available for the organism since no high energy phosphate compounds are formed. A detoxification mechanism for the metabolism of malic acid has been suggested. The possibility that some of the intermediary pyruvic acid may serve as a carbon source for the

TABLE XLVI

CHEMICAL ANALYSIS OF WINES AFTER MALO-LACTIC FERMENTATION

Inoculum	pH	Total Acid as Tartarate g/100 ml	Malic Acid g/100 ml	Volatile Acid as Acetate g/100 ml	Ethanol % v/v	Diacetyl + Acetoin ppm
Control, unfermented must, 23°Brix	3.5	0.75	0.338			
Control fermentation without bacteria	3.7	0.62	0.332	0.30	11.4	8.2
Lb. delbrueckii	3.9	0.43	tr	0.42	11.2	11.9
Lb. buchneri	3.9	0.43	tr	0.41	11.3	11.9
Lb. brevis	3.9	0.44	tr	0.36	11.5	4.9
Lc. citrovorum	3.9	0.44	tr	0.41	11.4	13.9
Pd. cerevisiae	3.9	0.43	tr	0.42	11.2	11.9

SOURCE: Adopted from Pilone et al. (1966).

NOTE: Analysis after completion of malo-lactic fermentation at 11-16°C, 1 to 2 months after inoculation.

growth of the lactic organism has also been considered in order to explain the stimulatory action of malic acid on growth (Kunkee, 1974).

Information on the chemical changes that take place during the malo-lactic fermentation has become available due to the extensive work by the Davis group (Pilone et al., 1966). These workers carried out a number of fermentations in the presence of selected cultures of lactic bacteria and determined various compounds employing conventional and GLC techniques (Table XLVI).

The changes in the composition of wines after the malo-lactic fermentation are striking. The practical disappearance of malic acid with a concomitant increase in pH, the significant increase in volatile acidity, the rise in diacetyl + acetoin values are more or less common to all the lactic bacteria employed. As expected, an increase in the content of ethyl lactate (not recorded in Table XLVI), up to 30 percent, has also been observed. It would be

TABLE XLVII

SENSORY ODOR THRESHOLDS IN GRAIN SPIRIT SOLUTIONS OF 9.4% (W/W)

Compounds	Odor Threshold *ppm*
Diacetyl	0.0025
Alcohols	
Hexylalcohol	5.2
Isoamyl alcohol	7.0
β-phenylethyl alcohol	7.5
Isobutyl alcohol	75.0
Esters	
Ethyl caprylate	0.25
Isoamyl acetate	0.20
β-phenylethyl acetate	0.65
Ethyl lactate	14
Ethyl acetate	17
Acids	
Isovaleric	0.7
Butyric	4.0
Isobutyric	8.1
Capric	8.2
Caproic	8.8
Caprylic	15.0
Propionic	20.0

SOURCE: After Salo (1970).

advisable at this point to note the flavor (odor) threshold levels of compounds appearing in alcoholic beverages as examined in alcoholic solution (Table XLVII).

In view of the odor thresholds of fermentation products, any change in the composition of wines will affect the bouquet qualities. The very low threshold of diacetyl requires some additional remarks. It has already been pointed out (see Chapter 3) that chemical analysis of diacetyl + acetoin have only limited value since the true concentrations of the important diacetyl may not be proportional to the total values of the combined determination. Dittrich and Kerner (1964) examined a number of German wines and found that in normal wines, diacetyl values were never above 0.6 mg/liter, whereas in defective wines with a lactic acid note, higher values up to 4.3 mg/liter were found (Table XLVIII).

Postel and Güvenc (1976) distinguish between normal values of diacetyl in German white wines (0.08 to 3.40 mg/liter) and those in red wines (0.26 to 4.06 mg/liter), the latter showing significantly higher values (average 1.46 vs. 0.42 mg/liter). That high concentrations of diacetyl may lead to the development of undesirable flavor defects such as "sauerkraut" is beyond dispute. On the other hand, there is little agreement about the maximum concentrations of diacetyl wines can tolerate. According to Rankine (1977), in Australian red table wines diacetyl up to 4.0 ppm should be considered a quality factor adding to the complexity of the wine. Most probably the tolerance of diacetyl in wines depends very much on the character of the wine and possibly also on the idiosyncrasy of the consumer.

The malo-lactic fermentation probably also affects the amino acid content of wines. According to Temperli and Künsch (1976), the total amount of free amino acids in fresh musts (variety *clevner*) after heating with skins fell from 6 to 7 g/liter to about 4 to 5 g/liter during the alcoholic fermentation. When the malo-lactic fermentation was carried out experimentally with a strain of *Lc. oenos,* a striking decrease in the arginine concentration (from 1858 to 59 mg/liter) was accompanied by an increase in ornithine (from 50 to 1430 mg/liter), indicating a vigorous arginine metabolism via the urea cycle. This reaction is not without effect on the flavor qualities of wine. Ornithine contrib-

TABLE XLVIII
DIACETYL IN WHITE WINES

	mg/liter
Austrian	0.5
Moselle	0.29
Rhine	0.57
Bordeaux	0.33
Burgundy	1.20
Algerian	0.73
Hungarian	0.72

SOURCE: Adapted from Ronkainen and Suomalainen (1969).

utes to a certain, hitherto undefined, flavor characteristic, which was considered preferable to arginine by all wine tasters.

Coming back to the malo-lactic fermentation, it is not surprising that the concentration of diacetyl is higher in wines that did undergo this fermentation and that such wines assume a distinctiveness and complexity that otherwise they would not have (Ingraham and Cooke, 1960). Data on the production of diacetyl by lactic acid bacteria during the fermentation of malic acid are rather scarce, and we do not know whether any difference exists between the hetero- and homolactic fermenters in this respect. On the other hand, values for diacetyl in normal wines that did not undergo the malo-lactic fermentation are quite low (Dittrich and Kerner, 1964), which would indicate that little diacetyl is formed during the alcoholic fermentation itself. This apparently is not the case. Suomalainen and Ronkainen (1968) have shown that yeasts synthesize α-acetolactic acid, which decomposes in the medium into diacetyl. However, in order to demonstrate substantial accumulation of diacetyl, it was necessary to remove the yeasts from the medium. In other words, yeasts seem to metabolize diacetyl resulting in minor amounts only. The suggestion by certain wine makers to improve the qualities of wines with a lactic acid defect by an additional fermentation with wine yeasts would be in accordance with the observations of these authors (Dittrich and Kerner, 1964).

The deacidification of wines by means of the malo-lactic fermentation under controlled conditions, which does not lead to the accumulation of by-products that may endanger wine quality, should be regarded as a desirable process for the improvement of the quality of acid wines (Castino et al., 1973). However, the very high acidities of such wines, which may have a pH as low as 3.0 in certain areas, is not suitable for the spontaneous development of the malo-lactic fermentation. Optimum pH for the lactic organisms is far above the pH of wine, but pH above 3.3 to 3.4, depending on the alcohol concentration and degree of sulfitation, may favor the fermentation. An additional obstacle to the malo-lactic fermentation has been ascribed to the bacteriostatic activity of bound SO_2 compounds, which are five to ten times less active than free SO_2 but may still be present at concentrations high enough for the inhibition of lactic bacteria at such low pH values (Lafon-Laforcade and Peynaud, 1974). The use of starter organisms in high acid wines should be helpful in order to start the process. The picture is totally different in the case of wines with lower acidities (and higher pH values). In wines with pH values above 3.8 (some authors believe this to be even lower), the malo-lactic fermentation may start without difficulty. This, however, is undesirable and may lead to the destruction of tartaric acid as well as malic acid, reducing the freshness of the wine which in some cases may result in bacterial spoilage altogether (Rankine et al., 1970). Hence, as long as malic acid is present in nonsterile wines, a biological instability due to a possible malo-lactic fermentation has to be taken into account. Some authors suggested the use of fumaric acid (0.05 to 0.15%) as an efficient inhibitor of the lactic bacteria. This should be added after the completion of the alcoholic fermentation and not to the grape juice, since fermenting yeasts apparently also metabolize fumaric acid (Rankine, 1977). Other authors believe that biological stability of low acid wines should be achieved by the controlled malo-lactic acid fermentation, employing suitable starters and environmental conditions. So far, too little is known about the flavor qualities of such wines. In countries where the addition of acids (malic, tartaric) is permitted (as in South Africa), local enologists do not view the malo-lactic fermentation as having a negative effect on the flavor qualities of their wines (van Wyk, 1976).

The degradation of tartaric acid by lactic organisms is not a frequent phenomenon but has been known since the early days of wine microbiology. Radler and Yanissis (1972), who studied the attack of many lactic bacteria on tartaric acid, describe only few strains of *Lb. plantarum* and *Lb. brevis* that were able to metabolish this substrate efficiently. Among the end products, CO_2, acetic acid, and lactic acid were found with the homofermenter, while succinate (and not lactate) was observed with *Lb. brevis*.

SULFUR IN WINES

The role of sulfur compounds in wine is not limited to the antiseptic and antioxidative properties of sulfur dioxide (SO_2). Other sulfur compounds may unfavorably affect the quality of wines, generally through the formation of hydrogen sulfide (H_2S) and other malodorant compounds (rotten eggs, *Weinböckser*) that must be eliminated in order to assure marketability. The production of such compounds are mediated microbiologically and will therefore be discussed in some detail.

In a very simplified scheme, the metabolism of sulfur compounds may be described as follows:

$$SO_4^= \rightarrow SO_3^= (SO_2) \rightarrow S° \rightarrow H_2S$$

Sulfates constitute the sulfur source for yeasts for the production of sulfur-containing amino acids and the synthesis of cell proteins and growth. Yeasts reduce sulfate via adenosine-5'-phosphosulfate (APS), 3'-phospho adenosine-5'-phosphosulfate (PAPS), and sulfite (Roy and Trudinger, 1970) (Fig. 26). Many yeasts reduce sulfates to a much greater extent than required for their biosynthetic activities, as a result of which sulfite, sulfur dioxide, and eventually hydrogen sulfide may accumulate. While H_2S may be swept out from the fermenting must with CO_2, it may yield a stale, rubbery, stagnant odor if it is not removed shortly after the fermentation ceases (Rankine, 1963). Considering the low threshold values of these compounds (below 1.0 ppm), even a slight accumulation of such volatile products may be detrimental to wine quality.

The sulfate content of musts ranges between 30 and 350 mg/liter (Amerine et al., 1967), providing ample substrate for the reducing activity of yeasts, which increases up to a sulfate level of 200 to 250 mg per liter (Würdig and Schlotter, 1971) as

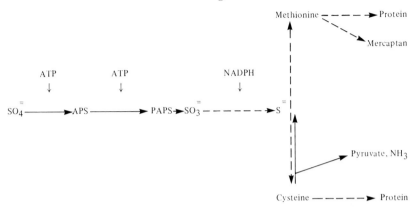

Figure 26. A Simplified Pathway of Sulfur in Wine Yeasts. After Eschenbruch (1974).

long as fermentable sugars are present (Dott et al., 1976). In addition, sulfur may reach the must from sulfur-containing fungicides used as spraying material in vineyards, from liquid sulfitation, or sulfur burning in the cellar. Also, sulfur-containing amino acids may lead to the production of the malodorous H_2S if no other sulfur-containing compounds are available (Rankine, 1963). Lower pH values and higher temperatures seem to favor H_2S formation.

Yeasts may vary a great deal in their sulfur-reducing potential (Table XLIX).

Cysteine may serve as a precursor for H_2S since yeasts can decompose it to H_2S, pyruvate, and NH_3. According to Lawrence and Cole (1968), an enzyme called cysteine desulfhydrase mediates the reduction of elemental, inorganic, and organic

TABLE XLIX

HYDROGEN SULFIDE FORMATION BY YEASTS IN PRESENCE OF VARIOUS SULFUR SOURCES (IN MG/LITER)

Yeast	None	Sulfur 30 ppm	$(NH_4)_2SO_4$ 125 ppm
S. cerevisiae, var. ellipsoideus	tr	20	6
Pichia membranefaciens	tr	14	2
S. uvarum	0	17	tr
H. anomala	0	31	5

SOURCE: Adapted from Hernadez (1964).

sulfur compounds to H_2S, thus showing poor substrate specificity. On the other hand, methionine, the other major sulfur-containing amino acid present in must, inhibits H_2S formation. Pantothenate and pyridoxine deficiencies in many strains of *S. cerevisiae* result in low levels of methionine in the cells due to insufficient formation of the intermediate acetyl homoserine (Wainwright, 1970). This leads to increased formation of H_2S. Certain metal ions stimulate the formation of hydrogen sulfide by yeasts. A significant increase of H_2S by copper-containing fungicides used in the vineyard before harvest has been observed. Zn- and Mg-containing fungicides may also have a similar effect (Eschenbruch, 1974). These ions may stimulate hydrogen sulfide formation during storage. The preference for the use of earthenware vessels over metal containers for the storage of wines in ancient civilizations may therefore be of considerable interest.

If either sulfate or sulfite is the only sulfur source, most yeast strains will produce little H_2S. If, on the other hand, colloidal sulfur is available in the medium, all strains produce large quantities of H_2S. Hence, most of the hydrogen sulfide in fermenting juices comes from residual sulfur. Attention should therefore be paid to the nature of fungicides used for spray in vineyards (Eschenbruch et al., 1978).

The first major step in sulfate reduction will lead to the formation of SO_2 (see Fig. 26). Not only different species but also different strains of the same yeast may vary in their reducing properties of sulfur compounds as can be seen in Table L.

Much higher values for the formation of sulfur dioxide were obtained in natural fermenting musts (100 ppm and above). These values varied considerably in different wine-producing localities (Würdig and Schlotter, 1971). With the isolation of pure cultures and the fermentation of individual yeasts derived from such natural fermentations, it became clear that certain natural yeasts may be extremely active in the reducing process. A yeast *(S. pastorianus)* was found to produce over 500 mg per liter SO_2. It may be concluded that the yeast population of naturally fermenting musts is very heterogenous with regard to the reduction of sulfur compounds, and one may speak of low and high SO_2-producing strains. The use of a suitable low producing

TABLE L
PRODUCTION OF SO₂ IN A SYNTHETIC MEDIUM (WICKERHAM) BY
VARIOUS YEAST STRAINS

Yeast		SO_2 mg/liter
S. bayanus	ML 8	9
	ML 1	23
	ML 16	15
S. cerevisiae	ML 14	13
	ML 35	4
	ML 42	18
	ML 58	4
	ML 93	18
S. florentinus	VR 16	6
	VR 19	8
	VR 20	8
	VR 26	10

SOURCE: Adapted from Delfini and Gaia (1977).

starter organism may be considered as a safeguard against the proliferation of high SO₂-producing organisms.

The regulation of SO₂ formation by wine yeasts has been studied by Eschenbruch et al. (1973). High concentrations of both cysteine and methionine seem to depress sulfate reduction and SO₂ accumulation. Sulfate reduction becomes no longer necessary for the synthesis of these amino acids. The fate of SO₂ in wines was studied with tracer techniques. When labelled SO₂ (150 mg/liter) was added prior to the fermentation, 30 percent of the label appeared as sulfate and about 10 percent in reduction compounds. Hence, part of the added SO₂ was oxidized while part was reduced. The difference between high and low sulfite-producing wine yeasts has been recently explained by an increase in the NADPH dependent sulfite reductase during the exponential growth phase of low sulfite strains (Dott and Trüper, 1978).

In addition to the formation of the malodorant H₂S in wines, *(Schwefelwasserstoffböckser)* the reduction of sulfur compounds may result in further objectionable metabolites (Tanner, 1969). Prolonged contact of wine with yeasts after the completion of the fermentation may lead to the formation of ethylmercaptan *(Lagerböckser):*

$$C_2H_5OH + H_2S \longrightarrow C_2H_5SH \quad \text{or} \quad 3CH_3CHO + 3H_2S \longrightarrow \left[CH_3C \overset{\overset{H}{|}}{\underset{S}{\diagdown}} \right]_3 \longrightarrow 3C_2H_5SH$$

TRI-THIOACETALDEHYDE   ETHYLMERCAPTAN

Ethylmercaptan has a strong garliclike odor and may undergo additional transformation to

$$2\ C_2H_5SH + CH_3CHO \xrightarrow{\text{Yeast}} CH_3\overset{\overset{\displaystyle SC_2H_5}{\diagup}}{\underset{\underset{\displaystyle H}{\diagdown}}{C}} - SC_2H_5 + H_2O$$

MERCAPTAL

or to

$$C_2H_5S - SC_2H_5$$

ETHYL DISULFIDE

Both compounds may be detrimental to wines because of their high boiling points which makes it more difficult to remove them by aeration (Tanner, 1969).

It is possible that sulfur-containing compounds may also affect favorably the flavor of wines. Schreier et al. (1975) described a number of sulfur-containing secondary amides, probably synthesized during the alcoholic fermentation. The occurrence of 3-(methylthio) propylamine in wine should be of importance since this was considered as a "flavor enhancer" (3-(Methylthio) Propylamine as a Flavor Enhancer, 1967).

MACERATION CARBONIQUE

To what extent does the technology of wine making affect the composition and activity of the microflora derived from the grapes, and in what way will this influence the flavor qualities of wines? These questions can be easily asked, but it is hard to give a reliable answer since a great number of experimental vinifications have to be carried out with suitable controls and statistical analysis.

An interesting example would be the fermentation of uncrushed grapes, first introduced in France (Flanzy, 1967) for the production of red wines. Grapes are filled into normal fermentation vats but are covered by gaseous carbon dioxide so that anaerobic conditions prevail from the very beginning of the fermentation. Grapes crushed during the handling and by their own weight free some of the must. A small percentage of the highly colored juice may be recovered while the majority of the grapes yield their must after being pressed and are fermented accordingly. A better retention of aroma and faster maturation is claimed for this process. Another interesting feature is the decrease in acidity by up to 65 percent of the malic acid content of the juice (Brechot et al., 1969). The change in the yeast flora during the *maceration carbonique* procedure was also examined. According to Brechot et al. (1962), there are fewer apiculate yeasts in this fermentation compared to the conventional technique, which of course may favorably affect wine quality. However, other authors could not confirm these results (Rosini and Fantozzi, 1975).

NOBLE ROT

Among the mycelial fungi, *Botrytis cinerea* occupies an important position in the wine industry. This fungus occurs on grapes in certain wine-growing areas of Europe as a result of natural infection (Figs. 27 and 28). Under suitable climatic conditions (rainy weather), the growth of this fungus on the surface of grapes in the form of a greyish (conidia) felt leads to the perforation of the skin and an increased loss of water through evaporation. This is more pronounced if sunny days follow the fungus infection. As a result of this, musts with higher sugar concentrations may be obtained, sometimes reaching even 40 percent, which makes them suitable for the production of sweet wines, e.g. Tokay. *Bo. cinerea,* also named "noble rot" *(pourriture noble, Edelfäule).* consumes relatively more acids than sugars, which also contributes to the sweetness of the must. The elevated sugar concentration, as well as the decrease in soluble nitrogen compounds (amino acids consumed by the fungus), reduce the rate of yeast growth and thus favor a less vigorous fermentation.

The degradation of acidity by *Bo. cinerea* is fairly selective. According to Peynaud et al. (1959), tartaric acid is more readily attacked by this fungus than malic acid. Dittrich et al. (1974, 1975) studied the effect of *Botrytis* infection on the quality of such

Flavor Microbiology

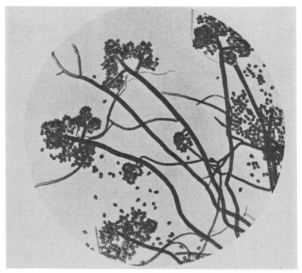

Figure 27. *Botrytis cinerea,* × 100. From G. Smith, *Industrial Mycology* (1960). With permission from Edward Arnold Publishers.

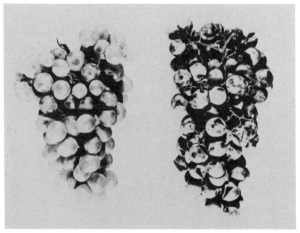

Figure 28. Nonbotrytized and Botrytized Semillon Grapes. From M. A. Amerine and M. A. Joslyn, *Table Wines, The Technology of Their Production* (1970). With permission of the University of California Press.

musts more extensively. They confirmed the preferential degradation of tartaric acid. Musts from infected grapes had 1 to 3 g/liter less tartaric acid than musts obtained from similar but uninfected grapes. Malic acid was consumed at a similar rate to sugars, resulting in a slight increase in musts from infected grapes. Interestingly, gluconic acid, which occurs in healthy grapes at minute concentrations, was found at a level of 4 g/liter in musts from infected grapes. Also, citric acid was slightly higher (about 200 mg/liter). High values for glycerol (up to 14.2 g/liter) were found in botrytized grapes, confirming earlier reports cited by Amerine (1954).

An increase in acetaldehyde, pyruvate, and ketoglutarate was also observed. The higher requirement for SO_2 in musts from botrytized grapes as a result of the accumulation of these SO_2 binding compounds was explained by the depletion of thiamine in infected grapes due to fungal metabolism. Normal grapes had about 250 ng/ml thiamine as compared to about 60 ng/ml in the juice of infected grapes. Although wine yeasts synthesize thiamine, the vigorous fermentation of musts probably requires additional thiamine, which in its absence causes the accumulation of these intermediate products of fermentation (Liebert, 1971).

The effect of botrytized grapes on the formation of fusel alcohols was studied in another paper (Dittrich and Sponholz, 1975). There was a significant increase in the content of higher alcohols of German wines produced from botrytized grapes but not at equal rates. While the concentration of isobutanol increased twofold, that of phenethanol was substantially inferior to that of wine made from healthy grapes. These results were not confirmed in French wines (Bertrand et al., 1976). Schreier and coworkers (1976) claimed that wines from botrytized grapes have lower concentrations of fusel oil alcohols. This was explained by the fact that the fungus metabolizes the amino acids of the must, thus decreasing the amount of precursors available for the formation of higher alcohols (Tokay-Ascu wines).

Many of the flavor compounds that appear in minute quantities in wine from botrytized grapes have not yet been examined sufficiently, but there is no doubt that this fungus has a definite effect on the flavor quality of such wines. Attempts have also been made to infect grapes artificially, before or after harvest

(Nelson and Amerine, 1957). The production of *Bo. cinerea* mycelium by a submerged culture process has been described by King et al. (1969). The effect of such mycelium added to fermenting grape juice has not yet been studied from an enological point of view.

SECONDARY FERMENTATIONS — YEAST

One of the most interesting cases in which a microbial process leads to the transformation of a fermented product into a beverage of higher organoleptic and commercial qualities is the *flor* type sherry. Wines with the highest reputation are those produced by a traditional process in the region of Jerez de la Frontera, Spain. There are two basic types: Oloroso sherry is produced from grapes with a high sugar content obtained after sun dehydration for a couple of days, yielding after fermentation 16 percent and over alcohol. This concentration usually prevents any secondary fermentation. The second type, Fino, has a lower alcohol content (14.5 to 15.5%), which permits the development of a film yeast, and a secondary fermentation takes place in the so-called solera system. There, in a number of barrels filled only to about 80 percent of their volume, the wine undergoes an oxidative process due to the metabolic activities of the so-called flor yeast. When the film is complete, it serves as a barrier, preventing direct contact of oxygen with the wine and promoting a reducing environment. Fresh wine is introduced into such a system, transferred from one layer of barrels to a successive one about six times until the characteristic flavor of the flor sherry is acquired. This may take five to six years or even longer. More detailed descriptions of the process may be found in textbooks on wine technology (Amerine et al., 1967; and others).

The traditional sherry from Jerez owes its distinctive character to a large number of factors: the palomino grapes grown on soils rich in limestone; the addition of *yeso* (gypsum) that leads to the formation of K_2SO_4, which imparts a somewhat bitter sensation; the fractional blending system of the solera; the oak barrels; and of course the nature of the yeast employed in the secondary fermentation. Recent advances in wine technology have introduced other methods of sherry production with the aim of eliminating the time-consuming and expensive solera system by either promoting the flor yeast under submerged culture conditions in suitable wines, with occasional agitation and aeration, or by the "baked process," eliminating altogether the microbial stage (Ough and Amerine, 1960).

There is little doubt that this very complicated wine with its distinctive aroma, as produced by the traditional solera system, differs from other wines mainly because of the activities of the flor yeasts. Opinions differ with regard to the speciation and correct nomenclature of this yeast. Earlier workers (Schanderl, 1959) believed that the flor yeast is but the oxidative stage of a strain of the normal wine yeast, which under suitable conditions assumes a more oxidative character. Other workers believe flor yeasts belong to special species such as *Saccharomyces beticus, S. bayanus, S. fermentati* or *S. prostoserdovii*, all somehow related to each other (Lodder, 1970).

According to Fornachon (1953), four stages may be distinguished in the growth and development of flor yeasts in an undisturbed vessel: (1) The yeast forms a complete film over the surface of the wine during the rapid growth phase. This is accompanied by a gradual dropping of the redox potential, the metabolism of ethanol to acetaldehyde, and the beginning of flor aroma formation. (2) The thick pellicle covers the wine, the liquid becomes more and more anaerobic with very low redox potentials. Aldehyde accumulation slows down. Some of the oxidized phenolic compounds are reduced, turning the dark amber color of the wine into a pale gold. (3) The film turns thinner, assumes a greyish tinge, and the redox potential starts to rise anew. Also, aldehydes are newly formed. (4) The film breaks down, and the wine is exposed to acetification.

The chemical transformations that the sherry wine undergoes during the oxidative stage are very complex and have been studied by a large number of workers. Webb and Noble (1976) enumerate over 100 volatile compounds that have been detected in sherries. The principal flavor components characteristic to sherry wine are aldehydes and acetals. Acetaldehyde is quantitatively the more important. During the solera process over 300 ppm may be formed (Marcilla et al., 1936). In the Californian sherry produced with the submerged culture of the flor yeast, much higher values of aldehydes (about 1000 ppm) were found

(Amerine, 1958). Considering the fact that normal wines contain only minor amounts of acetaldehyde, it is clear that these compounds contribute significantly to the formation of the sherry wine aroma. However, we still do not know the optimum concentration of aldehydes in sherry preferred by the consumer. A blending procedure for the production of sherry with conventional concentrations of acetaldehyde has been proposed by Webb et al. (1964) for the Californian type of submerged sherry.

Acetals are the product of condensation of aldehydes with two molecules of alcohols with the elimination of water:

$$CH_3CHO \ + \ 2 \ C_2H_5OH \longrightarrow \overset{\displaystyle OC_2H_5}{\underset{\displaystyle OC_2H_5}{CH_3\overset{|}{\underset{|}{CH}}}} \ + \ H_2O$$

ACETALDEHYDE   ETHANOL   DIETHYL ACETAL

Although there are indications of the occurrence of acetals in wine grapes and normal wines (Webb and Kepner, 1957; Lipis and Mamakowa, 1963), few quantitative data are available. However, it is assumed that during the solera process the amounts of the pleasantly smelling acetals are significantly increased (Webb et al., 1964). With the advent of gas chromatographic analysis, further insight into the qualitative nature of these acetals has been gained. Diethyl acetal was found in higher amounts in flor sherry than in the submerged product. Galetto et al. (1966) detected eight other acetals (ethyl active amyl, ethyl isoamyl, ethyl pentyl, isoamyl pentyl, diactive amyl, active amyl isoamyl 2-phenethyl, active amyl 2-phenethyl isoamyl acetal). Although it has been suggested that the organoleptic qualities of sherry wines be evaluated according to their acetal composition, there is little evidence that the acetals are formed by microbial activity since similar amounts of diethyl acetal could be demonstrated in "baked" sherries (Webb et al., 1964).

Both the flor sherry and the baked product contain acetoin in the range of 10 to 32 ppm (Guymon and Crowell, 1965). However, the submerged product is outstanding in its acetoin content, which may be as high as 350 ppm. Undoubtedly, the aerated culture leads to a higher metabolic activity of the flor yeast as compared to that of the solera system.

The effect of flor yeast on the accumulation and degradation of alcohols in the solera has been studied by various authors. Ethanol was found in some cases to decrease, which at least in part may be due to evaporation (Webb et al., 1966). Glycerol decreases considerably. Marcilla et al. (1936) claimed a 50 percent reduction of the glycerol concentration after completion of the solera system. Similar results were obtained in the submerged sherry (Amerine, 1958).

Headspace analysis of Australian flor sherries showed the aroma-significant compounds to be the ethyl esters of n-hexanoic, n-octanoic, n-decanoic acids as well as n-hexanol. After varying periods of flor film growth, changes in the composition of aroma substances took place. The concentration of ethyl n-octanoate and ethyl n-decanoate decreased while that of 2-phenethyl alcohol and diethylsuccinate increased (Williams and Strauss, 1978).

In addition to the customary fusel compounds, benzyl alcohol and 2-phenethyl alcohol were found in flor sherry of Spanish origin. Numerous publications deal with the occurrence of other alcohols, acids, and esters as determined by gas chromatography. Lactones, ethyl-4-hydroxybutyrate, ethyl pyroglutamate, diethyl succinate, diethyl malate, and 2-phenylethyl caproate were also identified. It is suspected that the presence of acetate and caproate esters of 2-phenethyl alcohol have a special place in the formation of the characteristic fino sherry aroma (Rodopulo et al., 1967). Augustyn et al. (1971) described the structure of a compound that is probably specific for the film sherry wines. This compound named solerone (after solera) has a typical winelike odor reminiscent of the odor of the cork of a bottle of quality Pinot Noir or resembling the odor that lingers in a wine glass several hours after an aged wine had been drunk from. It was described as 5-acetyldihydro-2(3H) furanone. It slowly decomposed at room temperature, yielding a product with little aroma. Infrared and mass spectrum analysis revealed a close relationship to the 4,5 dihydroxyhexanoic acid γ-lactone found in sherries. Its concentration seems to be a function of the length of time these wines were under film. It is speculated that solerone is formed first and later reduced to the corresponding hydroxylactone under the conditions prevailing below the film. The origin of solerone remains to be elucidated.

SOLERONE
5 - ACETYLDIHYDRO - 2(3H) FURANONE

4 - 5 DIHYDROXYHEXANOIC
ACID - γ - LACTONE

Flor sherries also contain other aroma compounds. The ethyl ester of 4-hydroxybutyric acid, which has a distinct scent, is in equilibrium with the relatively weakly scented γ-hydroxy-butyrolactone, which is quantitatively more important (Webb et al., 1967). Additional lactones, e.g. 4-ethoxy-4-hydroxybutyric acid γ-lactone, are known to occur in flor sherries and other wines, and pathways for their biosynthesis have been suggested (Muller et al., 1972, 1973).

Submerged culture sherries have also yielded some specific compounds. These are N-isopentyl and N-(2-phenethyl) acetamide. It was postulated that these compounds are formed by the reaction of isopentyl amine and 2-phenethylamine with active aldehyde or acetyl-CoA of the submerged culture (Webb et al., 1965).

Another secondary fermentation with large commercial implications is that employed for the production of sparkling wines, e.g. champagne. These wines are generally produced from dry wines, i.e. after the completion of the primary fermentation and the addition of fresh sugars (20 to 25 g/liter) and yeast for the second fermentation with the purpose of producing enough pressure in bottles or other containers in order to impart the sparkling character to the wine. The technology of champagne production has been reviewed by Bidan (1969). The high alcohol content (about 11%v) of the base wines necessitates a special strain of yeast that is capable of restarting the alcoholic fermentation under these conditions.

Although the main purpose of the second fermentation is to provide sufficient CO_2, there are also a number of changes in the flavor and bouquet of such wines. No doubt the autolysis of the yeast cells, which takes place during the fermentation at such elevated alcohol concentrations, must contribute to the character of these wines in addition to the sparkling nature. Bergner and Wagner (1965) examined the amino acid content of wines during the second fermentation and observed a significant increase in values of most amino acids. This may lead to increased levels of fusel oil alcohols in champagne (Konovalova et al., 1977). Other workers (Edelényi et al., 1976) claim that the changes in the composition of flavor components of sparkling wines are mainly quantitative. There was a significant decrease in isopentanol, whereas the content of isobutanol, isopropanol, and heptanol increased markedly during the aging of bottles after the secondary fermentation.

PHENOLIC COMPOUNDS

Among the minor components of must and wine, the phenolic compounds are very important from the point of view of flavor. Phenolic compounds that originate in grapes (mainly in seeds and grape skins) constitute a special group that comprises anthocyanins, flavonoids, volatile phenolics, and tannins. While white wines contain up to 500 mg/liter total phenol compounds, the concentration in red wines is much higher, 1200 to 1500 mg/liter, but may also exceed 2000 mg/liter. Obviously, many of the phenolic compounds are extracted from the grapes during the early phase of the fermentation and before separation of the must from the skins (Singleton and Noble, 1976).

Phenolic compounds affect the flavor of wine in a number of ways. Most important is the astringency of wines for which various polyphenol compounds (tannins) are responsible. These compounds also contribute to the bitter taste of wines. Other phenols, e.g. tyrosol $C_6H_4(OH)CH_2CH_2OH$, hydroxycinnamates $C_6H_4(OH)CH=CH_2COOH$), which occur near or above the taste threshold levels in wines, are also bitter compounds. Among the volatile phenols, cresols, guaiacols, vanillin, etc. appear below threshold levels in most wines but are probably additive to yield a sensory effect as a group (5 to 10 mg/liter). Tyrosol may reach even higher values (up to 45 mg/liter) and due to its agreeable odor, described as honeylike, waxy, or old fruit, may certainly affect the bouquet of wines.

During wine fermentation, wine yeasts may convert some of the nonphenolic precursors to phenolic compounds (Stevens, 1961). However, this subject has not yet been sufficiently documented. On the other hand, alcohol produced during the fermentation on the skins will significantly affect the extraction of the plant phenolic substances. The concentration of these compounds will decrease during the vinification processes due to a number of factors. Yeast may adsorb and precipitate certain phenols. Assuming a production of yeast biomass of about 10 g (wet weight) per liter of wine, the potential removal of 1000 mg/liter of tannins by yeast adsorption may well explain the drop in red wine total phenols from the predicted total extractable phenols of red grape musts. In fact, wine prepared by the addition of alcohol to a must had about 20 percent higher total phenols than wine prepared from the same must by fermentation (Berg and Akiyoshi, 1957). Some of the phenols may be assimilated by yeasts. There is good evidence that yeasts may be adapted to efficient utilization of phenol compounds as carbon source. *S. cerevisiae* reduced the phenol content in a medium from about 500 mg/liter to practically zero. *Torulopsis utilis* was adapted to grow in a medium containing 5000 mg/liter of a mixture of phenols containing cresols (Zimmerman, 1958; Rieche, 1962). Also, *Botrytis cinerea,* the noble mold, seems to take part in the reduction of the phenol content of infected grapes. Mold enzymes attack anthocyanins more actively than tannins (Bolcato et al., 1964). Anthocyanins are apparently also reduced during the malo-lactic fermentation by a *Leuconostoc* species, resulting in a decreased color of the wine (Vetsch and Lüthi, 1964).

Another way for the reduction of phenolic compounds may be related to the formation of aldehydes during the fermentation. Cantarelli (1958) has shown that the high phenol and anthocyanin content of red wines is converted rapidly to discolored and insoluble products by reaction of phenols with aldehydes produced by the aerobic yeasts.

Practically every phenolic derivative occurring in musts possesses some ability to inhibit microbial growth. Although this is usually not a powerful inhibition, under certain conditions of stress, e.g. in the presence of alcohol, this inhibition may become of practical importance. In fact, highly colored grape varieties may become difficult to ferment to dryness owing to their inhibition of yeasts by anthocyanins and tannins of the must. Since red wines usually undergo a malo-lactic fermentation, this is another indication that these compounds little affect the lactic acid bacteria (compare with Peynaud and Dupuy, 1964). On the other hand, in the case of secondary fermentations that begin with a comparatively high alcohol content (about 10%), difficulties may be encountered in the completion of such fermentations if red wine bases are employed. Caffeic acid (3.4,-dihydroxycinnamic acid) is probably the most inhibitory to wine yeasts (Epernay strain). As wine ages, the amount of caffeic acid increases, which makes the secondary fermentation of older base wines even more difficult (Sikovec, 1966). Apiculate yeasts, as well as the aerobic stage of the flor yeasts, seem to be more susceptible to phenolic compounds (Schanderl, 1962).

For an exhaustive review on the effect of phenolic substances on flavor formation in wines, see Singleton and Esau (1969).

FLAVOR DEFECTS IN WINES

Drastic changes in the composition of a certain wine, e.g. a sharp increase in acidity, are generally considered as spoilage and will not be discussed here. Slight changes that do not drastically affect the nature of a wine but may contribute a certain undesirable flavor deviation will be mentioned.

The use of the unsaturated fatty acid, sorbic acid (2,4-hexadienoic acid), as an antimicrobial agent against yeast and aerobic bacteria in wines is now fairly widespread. Most of the people who consume these wines are not able to detect this preservative at the concentration usually employed (100 to 200 ppm). However, a flavor defect known as *geranium flavor* may develop as a result of the use of this preservative in wines if not sufficiently protected by SO_2. Wines to which sorbic acid was added during certain stages of the vinification process were found to be liable to such a flavor defect.

The geranium flavor reminds one of the leaves of the pelargonium plant. The chemical compound responsible for this odor is apparently 2,4-hexadien-1-ol. Some authors believe this to be 2-ethoxyhexa 3,5-diene (Crowell and Guymon, 1975). This latter compound is probably derived from hexadienol. After Crowell and Guymon (1975):

$$CH_3-CH=CH-CH=CH-COOH \qquad (2,4\text{-hexadienoic acid})$$

sorbic acid

$$\downarrow \text{Lactic bacteria}$$

$$CH_3-CH=CH-CH=CH-CH_2OH \qquad (2,4\text{-hexadiene-1-ol})$$

sorbyl alcohol

$$\downarrow \text{H+}$$

$$CH_3-CH(OH)-CH=CH-CH=CH_2 \qquad (3,5\text{-hexadien-2-ol})$$

$$\downarrow \text{Ethanol}$$

$$CH_3-CH(OC_2H_5)-CH=CH-CH=CH_2, \qquad (2\text{-ethoxyhexa }3,5\text{-diene})$$

The involvement of lactic acid bacteria in the transformation of sorbic acid into the corresponding alcohol was suspected. In a series of controlled experiments with different lactic organisms, Radler (1976) observed that 2,4-hexadienol was formed by most of the heterolactic bacteria including *Lc. oenos, Lb. buchneri,* and *Lb. hilgardii* but rarely by homolactic organisms. According to Heintze (1975), the geranium defect of wines is the result of the interaction of SO_2, lactic acid, ethanol, and sorbic acid, leading to the formation of a sulfonated hexa 2.4-dien-1-ol lactate. (For the metabolism of sorbic acid by mycelial fungi, see Tahara et al., 1977.)

The formation of the geranium odor as a result of the reduction of sorbic acid does not result in the exhaustion of the preservative since minute quantities (a few ppm) of the odor may already be detected by smelling. The use of sorbic acid cannot, therefore, be considered safe unless the activity of lactic acid bacteria can be suitably controlled. Other flavor defects in fermented beverages will be discussed in Chapter 8.

OTHER FRUIT JUICES

Comparatively little has been published on the production of flavor in nongrape alcoholic beverages. Apple cider is the best known example of such fruit juice that undergoes an alcoholic fermentation (5 to 8% v/v alcohol). Pollard et al. (1966) reviewed the flavor components of apple cider and pointed out that the most important change during the processing is the transformation of the fruity acid character of the juice to the softer mature flavor of the cider. It had been shown earlier (Whiting and Coggins, 1960) that these changes were due partly to the bacterial activities during or after the yeast fermentation. Lactic acid bacteria may attack sugars in nonsulfited juices with pH above 3.8. Concurrently, malic acid is converted to lactic acid in a manner similar to the malo-lactic fermentation in wines. Also, the yeast fermentation contributes to the general increase in cider acidity due to the significant formation of succinic acid. This is obscured in the grape juice fermentation owing to the changes in the tartaric acidity in the CO_2-saturated environment. There was little difference between the yeast strains employed for cider making (Thoukis et al., 1965).

Another component of microbial origin associated with the desirable flavor of cider is the relatively high level of fusel oil. Although some of these higher alcohols may be present in certain varieties of apple juices, the bulk of this fraction is derived from the yeast fermentation. In comparing the content of higher alcohols in apple juice and cider, it was found that with the exception of n-butanol, alcohols increase significantly during the fermentation, primarily isobutyl, iso and active amyl alcohols, as well as 2-phenyl-ethanol. In a series of laboratory experiments, it was shown that the fermentation of apple juice by its natural microflora usually resulted in a much higher fusel oil content than when the fermentation was carried out with a culture yeast used in this process: 197 to 335 ppm versus 151 to 167 ppm respectively; in the case of 2-phenyl ethanol, 127 to 254 ppm as against 33 to 65 ppm with a culture yeast. Indeed, the natural flora of individual apple yeasts vary a great deal in their ability to produce higher alcohols. However, single cultures of the natural flora do not seem to produce as much phenyl ethanol as do mixed cultures (Table LI).

Usually the fusel oil content of ciders are intermediate between that of low gravity beers and wines, which would correlate with their respective ethanol content. However, in the case of cider, the scented character of some of these alcohols may constitute an important part of its aroma. Taste panels have shown that up to 200 ppm of higher alcohols were considered essential to typical cider aroma (Pollard et al., 1966).

In summing up this section on the fermentation of musts of grapes and other fruit juices, it has to be pointed out that in spite of the tremendous analytical work that has been done in the isolation and characterization of flavor components of these

TABLE LI

EFFECT OF INDIVIDUAL YEASTS ON FUSEL OIL PRODUCTION
IN APPLE JUICE

	Fermentation Product		
Yeast	Higher aliphatic alcohols (ppm)	2-phenyl-ethanol (ppm)	Ethanol (% v/v)
Juice yeast:			
Hansenula sp.	186	27	4.8
Kloeckera apiculata	195	24	5.0
K. magna	49	2	3.2
Candida pulcherrima	153	14	4.8
C. parapsilosis	119	12	2.7
Torulopsis datilla	76	21	5.0
Culture yeast:			
Saccharomyces cerevisiae	155	9	5.1
S. delbrueckii	157	13	5.0
S. uvarum	159	53	5.0
Burgundy	171	20	4.9
Port 350 R	155	7	5.1

SOURCE: Adapted from Pollard et al. (1966).

products, we are still a long way from the complete description of the flavor profile of a certain type or brand of wine. However, there is little doubt that microbes, both yeast and bacteria, play a very important role in the formation and development of the flavor characteristics of such products.

BIBLIOGRAPHY

Amerine, M. A.: Composition of wines. I. Organic constituents. *Adv Food Res, 5:*353, 1954.

Amerine, M. A.: Acetaldehyde formation in submerged cultures of *Saccharomyces beticus. Appl Microbiol, 6:*160, 1958.

Amerine, M. A., Berg, H. W., and Cruess, W. V.: *The Technology of Wine Making,* 2d ed. Westport, Connecticut, Avi, 1967.

Augustyn, O. P. H., Van Wyck, C. J., Muller, C. J., Kepner, R. E. and Webb, A. D.: The structure of solerone, a substituted γ-lactone involved in wine aroma. *J Agric Food Chem, 19:*1128, 1971.

Benda, I. and Schmitt, A.: Enological studies on biological acid decomposition in musts by *Schizosaccharomyces pombe. Weinberg u Keller, 13:*239, 1966.

Berg, H. W. and Akiyoshi, M.: The effect of various must treatments on the color and tannin content of red grape juices. *Food Res, 4:*373, 1957.

Bergner, K. G. and Wagner, H.: Die Freien Aminosäuren während der Flaschen-und Tankgärung von Sekt. *Mitt Rebe Wein, 15:*181, 1965.

Bertrand, A., Pissard, R., Sarre, C., and Sapis, J. C.: Etude de l'influence de la pourriture grise de raisins *(Botrytis cinerea)* sur la composition chimique et la qualitee des vins. *Connais Vigne Vin, 10:*427, 1976.

Bertuccioli, M. and Viani, R.: Red wine aroma: identification of headspace constituents. *J Sci Food Agric, 27:*1035, 1976.

Bidan, P.: Methods of preparation and characteristics of sparkling wines. *Bull Office Inter Vigne Vin, 42:*34, 1961.

Bolcato, V., Lamparelli, F., and Losito, F.: Azione della *B. cinerea* e del lievito sulle sustanze coloranti dei mosti d'uva. *Riv Viticolt Enol, 17:*415, 1964.

Brechot, P., Chauvet, J., Croson, M., and Irmann, R.: The metabolism of malic acid during vinification. *Ann Technol Agric, 18:*293, 1969.

Brechot, P., Chauvet, J., and Girard, H.: Identification of yeasts in a Beaujolais must during fermentation. *Ann Technol Agric, 11:*235, 1962.

Cantarelli, C.: Invecchiamento naturale lento ed invecchiamento accelerato del vino in rapporto ai charatteri organolettici ed al valore biologico del prodotto. *Accad Ital Vite Vino Siena Atti, 10:*432, 1958.

Cappucci, C.: Osservazione sulle probabili cause della inconsueta acidita volatile di alcuni vini della regione Emiliano-Romagnola. *Riv Vitticolt Enol, 1:*386, 1948.

Castino, M., Usseglio-Tomasset, L., and Gandini, A.: *Factors which affect the spontaneous initiation of the malo lactic fermentation in wines.* 4th Long Ashton Symposium, Bristol, 1973.

Cerutti, G., Gelati, R., and Zappavigna, R.: Fermentzione del mosto d'uva in presenza di aminoacidi. *Riv Viticolt Enol, 31:*3, 1978.

Chen, E. C. H.: Keto acid decarboxylase and alcohol dehydrogenase activities of yeast in relation to the formation of fusel alcohols. *Can Inst Food Sci Technol J, 10:*27, 1977.

Crowell, E. A. and Guymon, J. F.: Studies of caprylic, capric, lauric, and other free fatty acids in brandies by gas chromatography. *Am J Enol Viticult, 20:*155, 1969.

————: Wine constituents arising from sorbic acid addition and identification of 2-ethoxyhexa-3,5 diene as source of geranium-like off-odor. *Am J Enol Viticult, 26:*97, 1975.

Delfini, C. and Gaia, P.: Indagine sulla produzione di anidride solforosa nel corso della fermentazione alcolica. *Vini d'Italia, 19:*239, 1977.

Dittrich, H. H. and Kerner, E.: Diacetyl als Weinfehler Ursache und Beseitigung des "Milchsäuretones," *Weinwissensch, 19:*528, 1964.

Dittrich, H. H. and Sponholz, W. R.: Die Aminosäureabnahme in Botrytis-infizierten Traubenbeeren und die Bildung Höhere Alkohole in diesen Mosten bei ihrer Vergärung. *Weinwissensch, 30:*188, 1975.

Dittrich, H. H., Sponholz, W. R., and Göbel, H. G.: Vergleichende Untersuchungen von Mosten und Weinen aus gesunden und aus Botrytis-infizierten Traubenbeeren. *Vitis, 13:*336, 1975.

Dittrich, H. H., Sponholz, W. R., and Kost, W.: Vergleichende Untersuchungen von Mosten und Weinen aus Botrytis-infizierten Traubenbeeren. *Vitis, 13:*36, 1974.

Dittrich, H. H. and Wenzel, K.: Die Abhängigkeit der Schaumbildung bei der Gärung von der Hefe und von der Most-Behandlung. *Weinwissensch, 31:*263, 1976.

Domerq, S.: Classification of wine grapes in the Gironde. *Ann Technol Agric, 6:*5, 139, 1965.

Dott, W., Heinzel, M., and Trüper, H. G.: Sulfite formation by wine yeasts. I. Relationships between growth, fermentation and sulfite formation. *Arch Mikrobiol, 107:*289, 1976.

Dott, W. and Trüper, H. G.: Sulfite formation by wine yeasts. VI. Regulation of biosynthesis of NADPH and BV-dependent sulfite reductases. *Arch Mikrobiol, 118:*249, 1978.

Drawert, F. and Rapp, A.: Gas-Chromatographische Untersuchung Pflanzlicher Aromen. *Chromatographia, 2:*446, 1969.

Drawert, F., Rapp, A., and Ulrich, W.: Über Inhaltsstoffe von Mosten und Weinen. *Vitis, 5:*195, 1965.

Drzaga, B., Swajdo, Z., and Gasik, A.: Effect of sulfitation of wine must on the contents of glycerol and non sugar extract in the wine. *Pzemysh Spozywecy, 30:*439, 1976.

Edelényi, M., Tóth, Á, and Szabo, M. T.: Gaschromatographische Untersuchung der Geruchs und Geschmackstoffe von Sekt. Die Nährung, *20:*347, 1976.

Ehrlich, F.: Conditions for the formation of fusel oil and its connection with protein synthesis. *Ber, 40:*1027, 1907.

Eschenbruch, R.: Sulfite and sulfide formation during winemaking. A Review. *Am J Enol Viticult, 25:*151, 1974.

Eschenbruch, R., Bonish, P., and Fisher, B. M.: The production of H_2S by pure culture wine yeasts. *Vitis, 17:*67, 1978.

Eschenbruch, R., Haasbroeck, F. J., and Villiers, J. F.: On the metabolism of sulphate and sulphite during the fermentation of grape must by *Saccharomyces cerevisiae. Arch Mikrobiol, 93:*259, 1973.

Ferreira, J. D.: The growth and fermentation characteristics of six yeast strains of several temperatures. *Am J Enol Viticult, 10:*1, 1959.

Flanzy, M.: Quelques nouveaux procédés de vinification. *Bull Off Int Vigne Vin, 40:*283, 1967.

Fornachon, J. C. M.: Studies on the Sherry Flor. Australian Wine Board, Adelaide, 1953.

Galetto, W. G., Webb, A. D., and Kepner, R. E.: Identification of some acetals in an extract of submerged-culture flor sherry. *Am J Enol Viticult, 17:*11, 1966.

Garvie, E. I.: *Leuconostoc oenos sp. nov. J Gen Microbiol, 48:*431, 1967.

Genevois, L.: Acide succinique et glycerine dans la fermentation alcoholique. *Bull Soc Chim Biol, 18:*295, 1936.

Goranow, N.: Dynamik der Bildugn von Aromastoffen während der alkoholischen Gärung von Traubenmost. *Mitt Rebe Wein Obst Fruchtverw, 26:*5, 1976.

Goto, S., Yamazaki, Y., Yamakawa, Y., and Yokotsuka, I.: Decomposition of malic acid in grape must by wine and wild yeasts. *Hakkogaku, 56:*133, 1978.

Goto, S. and Yokotsuka, I.: Wild yeast population in fresh grape musts of different harvest times. *J Ferment Technol, 55:*451, 1977.

Greenshields, R. N.: Acetic acid, vinegar. In Rose, A. H. (Ed.): *Economic Microbiology.* London, New York, San Francisco, Acad Pr, 1978.

Guymon, J. F.: Mutant strains of *Saccharomyces cerevisiae* applied to studies of higher alcohol formation. *Dev Ind Microbiol, 7:*88, 1966.

Guymon, J. F. and Crowell, E. A.: The formation of acetoin and diacetyl during fermentation, and the levels found in wine. *Am J Enol Viticult, 16:*85, 1965.

Guymon, J. F. and Heitz, J. E.: The fusel oil content of California wines. *Food Technol, 6:*359, 1952.

Guymon, J. F., Ingraham, J. L., and Crowell, E. A.: Influence of aeration upon the formation of higher alcohols by yeasts. *Am J Enol Viticult, 12:*60, 1961.

Harvalia, A., Donilatos, N., and Vassiliou, M.: Relation entre la teneur de vins en alcools superieurs et la teneur des mouts en substances azotees. *Bull de L'O.I.V., 541:*222, 1976.

Heintze, K.: Über die Bildung des sogenannten Geranientones in Wein. *Weinberg u Keller, 22:*335, 1975.

Hernandez, M. R.: Production of H₂S by wine yeasts grown with several sulfur containing compounds. *Sem Vitivinicola, 19:*2359, 1964.

Hieke, E. and Vollbrecht, D.: Zur Bildung von Butanol-2 durch Lactobacillen und Hefen. *Arch Microbiol, 99:*345, 1974.

Hinreiner, E., Filipello, F., Berg, H. W., and Webb, A. D.: Evaluation of thresholds and minimum difference concentrations for various constituents of wines. *Food Technol, 9:*489, 1955.

Huglin, M. P., Meyer, J. P., and Schaeffer, A.: Elaboration de concentrés des mouts de raising peu acides a l'aide de levures du genre *Schizosaccharomyces. Acad Agric France,* 1976, p. 412.

Ingraham, J. L. and Cooke, G. M.: A survey of the incidence of malo-lactic fermentation in California table wines. *Am J Enol Viticult, 11:*160, 1960.

King, A. D., Camirand, W. M., and Mihara, K. L.: Submerged-culture production of *Botrytis cinerea* mycelium. *Am J Enol Viticult, 20:*146, 1969.

Konovalova, L. A., Dzhurikyants, N. G., and Goryaev, M. I.: Formation of higher alcohols in wine. *Vinodelic Vinogradarstvo SSSR, 2:*14, 1977.

Kunkee, R. E.: Malo-lactic fermentation and winemaking. In Webb, A. D. (Ed.): *Chemistry of Winemaking.* Washington, Am Chemical, 1974.

Kunkee, R. E. and Amerine, M. A.: Yeasts in wine-making. In Rose, A. H. and Harrison, J. S. (Eds.): *The Yeast,* vol. 3. London & New York, Acad Pr, 1970.

Lafon-Lafourcade, S. and Peynaud, E.: Sur les taux des acides cetonigues formes au cours de la fermentation alcoolique. *Ann Inst Pasteur, 110:*766, 1966.

———: Sur l'action antibacterienne de l'anydrida sulfureux sous forme libre et sous forme combinee. *Conn Vigne Vin, 8:*187, 1974.

Lawrence, W. C. and Cole, E. R.: Yeast sulfur metabolism and the formation by hydrogen sulfide in brewing fermentations. *Wallerstein Comm, 31:*95, 1968.

Liebert, H. P.: Der Einfluss von Apfel und Weinsaure sowie von Vitamin B₁ auf die Glycerinbildung bei der alkoholischen Gärung. *Zentralbl Bakteriol, 126:*162, 1971.

Lipis, B. V. and Mamakowa, Z. A.: Analysis of volatile components of wine, cognac and spirits by gas liquid chromatography. *Vinodeliei Vinogradorstvo SSSR, 23:*7, 1963.

Lodder, J.: *The Yeasts: A Taxonomic Study,* 2d ed. Amsterdam, North Holland Pub, 1970.

Lüthi, H. and Vetsch, U.: Contributions to knowledge of the malo-lactic fermentation in wines and ciders. *J Appl Bacteriol, 22:*384, 1959.

Malan, C. E.: Taste properties of sterile musts fermented by pure yeasts. *Riv Viticol Enol, 9:*11, 1956.

Marcilla, J. A., Alas, G., and Feduchy, E.: Contribucion al estudio de la levaduras que forman velo sobre ciertos vinos de elevado grado alcoholico. *Ann Centro Invest Vinicolos, 1:*1, 1936.

Muller, C. J., Kepner, R. E., and Webb, A. D.: Lactones in wines, a review. *Am J Enol Viticult, 24:*5, 1970.

———: Identification of 4-ethoxy-4-hydroxy butyric acid γ-lactone (5-ethoxy dihydro-2(3H)-furanone) as an aroma component of wine from *Vitis vinifera* var. Ruby Cabernet. *J Agric Food Chem, 20:*193, 1972.

Nelson, K. E. and Amerine, M. A.: The use of *Botrytis cinerea* Pers. in the production of sweet table wines. *Hilgardia, 26:*521, 1957.

Neubauer, O. and Fromherz, K.: Über den Abbau der Aminosauren bei der Hefegärung. *Z Phys Chem, 70:*326, 1911.

Ough, C. S., Guymon, J. F., and Crowell, E. A.: Formation of higher alcohols during grape juice fermentations at various temperatures. *J Food Sci, 31:*620, 1966.

Oura, E.: Reaction products of yeast fermentations. *Proc Biochem, 12:*19, 1977.

Park, Y. H. and Bertrand, A.: Contribution a l'etude des levures des cognac. *Conn Vigne Vin, 8:*343, 1974.

Peynaud, E. and Dupuy, P.: Methodes d'isolement, de culture, et de classification des bacteries malo lactiques. *Bull Office Intern Vigne Vin, 37:*908, 1964.

Peynaud, E. and Guiberteau, G.: Sur la formation des alcools superieure par les levures de vinification. *Ann Technol Agric, 11:*85, 1962.

Peynaud, E., Lafourcade, S., and Charpeutié, Y.: Recherches sur les transformations du raisin par *Botrytis cinerea. Vignes et Vins, 75:*6, 1959.

Peynaud, E., Lafon-Lafourcade, S., and Guimberteau, G.: L(+) lactic acid and D(−) lactic acids in wines. *Am J Enol Viticult, 17:*302, 1966.

———: Nature de l'acide lactique forme par les levures — un charactere specifique de *Saccharomyces veronae* Lodder et Van Rij. *Antonie Van Leeuwenhoek J Microbiol Serol, 33:*49, 1967.

Philip, T. and Nelson, F. E.: A procedure for quantitative estimation of malic and tartaric acids of grape juice. *J Food Sci, 38:*18, 1973.

Pilone, G. J., Kunkee, R. E., and Webb, A. D.: Chemical characterization of wines fermented with various malo lactic bacteria. *Appl Microbiol, 14:*608, 1966.

Pollard, A., Kieser, M. E., and Beech, F. W.: Factors influencing the flavor of cider: the effect of fermentation treatments on fusel oil production. *J Appl Bacteriol, 29:*253, 1966.

Postel, W. and Güvenc, U.: Gaschromatographische Bestimmung von Diacetyl, Acetoin und 2,3-Pentadion in Wein. *Z Lebensm Unters Forsch, 161:*35, 1976.

Radler, F.: Bedeutung und Möglichkeiten der Verwendung von Reinkulturen von Hefen bei der Weinbereitung. *Weinberg u Keller, 20:*339, 1973.

———: Degradation de l'acide sorbique par les bacteries. *Bull Office Intern Vigne Vin, 49:*629, 1976.

Radler, F., Baranowski, K., and Kuczynski, J. T.: *Malic Enzyme of Saccharomyces.* 5th International Fermentation Symposium, 1977.

Radler, F. and Yannissis, C.: Weinsäureabbau bei Milchsäurebakterien. *Arch Mikrobiol, 82:*219, 1972.

Ramos, C. L. and Gomes, G. L.: Higher alcohols in port wine. *Anais do Inst Vinho do Porto, 25:*95, 1974.

Rankine, B. C.: Nature, origine and prevention of hydrogen sulfide aroma in wines. *J Sci Food Agric, 14:*79, 1963.

———: Formation of higher alcohols by wine yeasts, and relation to taste thresholds. *J Sci Food Agric, 18:*583, 1967.

———: Formation of α-ketoglutaric acid by wine yeasts and its enological significance. *J Sci Food Agric, 19:*624, 1968.

———: Developments in malo-lactic fermentation of Australian red table wines. *Am J Enol Viticult, 28:*27, 1977.

Rankine, B. C., Fornachon, J. C. M., Bridson, D. A., and Cellier, K. M.: Malo-lactic fermentation in Australian dry red wines. *J Sci Food Agric, 21:*471, 1970.

Rankine, B. C. and Lloyd, B.: Quantitative assessment of dominance of added yeast in wine fermentations. *J Sci Food Agric, 11:*793, 1963.

Rankine, B. C. and Pocock, K. F.: β-phenylethanol and hexanol in wines. *Vitis, 8:*23, 1969.

Ribereau-Gayon, J. and Peynaud, E.: Sur la formation des acides acetique, lactique et citrique au cours de la fermentation alcoolique. *Compt Rend, 222:*457, 1946.

———: *Traite d'Oenologie,* vol. 2. Paris, Librarie Polytech Ch Beranger, 1961.

———: *Traite d'Oenologie,* vol. 1. Paris, Librarie Polytech Ch Beranger, 1964.

Rieche, A., Hilgetag, G., Martini, A., and Lorenz, M.: *Continuous production of yeast from phenol-containing substrates.* Continuous Cultivation Microorganisms Symposium. Prague, 2:*293, 1962.

Rodopulo, A. K., Levchenko, T. N., Pisarnitskiy, A. F., Bezzubov, A. A., and Martakov, A. A.: Issledovanie biochimicheskich processov pry cheresovanie vina. *Prikl Biokhim Mikrobiol, 3:*614, 1967.

Ronkainen, P. and Suomalainen, H.: Die Bestimmung vicinaler Diketone in Weissweine unter Anwendung der Dampfraum-analyse. *Mitt Rebe Wein, 19:*102, 1969.

Rosini, G.: Influenza del sistema di vinificazione sulla flora blastomicetica. *Vini d'Italia, 17:*321, 1975.

Rosini, G. and Fantozzi, P.: Applicazione della macerazione carbonica alla produzione del vino Rosso Piceno. *Vini d'Italia, 17:*515, 1975.

Roy, A. B. and Trudinger, P. A.: *The Biochemistry of Inorganic Compounds of Sulphur.* Cambridge, Massachusetts, Cambridge U Pr, 1970.

Salo, P.: Determining the odor thresholds for some compounds in alcoholic beverages. *J Food Sci, 35:*95, 1970.

Sapis, J. C. and Ribereau-Gayon, P.: A study of tyrosol, tryptophol, phenylethyl alcohol, and γ-butyrolactone, secondary products of the alcoholic fermentation in wines. *Ann Technol Agric, 18:*221, 1969.

Sapis-Domercq, S., Bertrand, A., Mur, F., and Sarre, C.: Influence des produits de traitement de la vigne sur la microflore levurine. *Conn Vigne Vin, 10:*369, 1976.

Schanderl, H.: Der Einfluss von Polyphenolen und Gerbstoffe auf die Physiologie der Weinhefe und der Wert des pH-7 Tests für die Auswahl von Sektgrundweinen. *Mitt, 12A:*265, 1962.

Schanderl, H.: *Die Mikrobiologie des Mostes und Weines,* vol. 2, Stuttgart, Ulmer Verlag, 1959.

Schreier, P., Drawert, F., and Junker, A.: Gaschromatographisch massenspektrometrische Untersuchung flüchtige Inhaltsstoffe des Weines. IV. Nachweis Sekundäre Amide in Wein. *Z Lebensm Unters Forsch, 157:*34, 1975.

————: GLC determination of constituents of fermented beverages. X. Qualitative determination of trace aroma components (μg/lt) in wines. *Chem Mikrobiol Technol Lebensm, 5:*45, 1977.

Schreier, P., Drawert, F., Kerenyi, A., and Junker, A.: Gaschromatographisch massenspectrometrische Untersuchung flüchtiger Inhaltsstoffe des Weines. VI. Aromastoffe in Tokajer Trockenbeereauslese (Aszo)-Weinen. *Z Lebensm Unters Forsch, 161:*249, 1976.

SentheShanmugathan, S.: The mechanism of formation of higher alcohols from amino acids by *Saccharomyces cerevisiae. Biochem J, 74:*568, 1960.

Sikovec, S.: Der Einfluss einigen Polyphenole auf die Physiologie von Weinhefen. I. Der Einfluss von Polyphenolen auf den Verlauf der Alkoholischen Gärung insbesonders von Umgärungen. *Mitt, 16:*127, 1966.

Singh, R. and Kunkee, R. E.: Alcohol dehydrogenase activities of wine yeasts in relation to higher alcohol formation. *Appl Environ Microbiol, 32:*666, 1976.

Singleton, V. L. and Esau, P.: *Phenolic Substances in Grapes and Wine, and their Significance. Adv Food Res Suppl,* vol. 1, New York and London, Acad Pr, 1969.

Singleton, V. L. and Noble, A. C.: *Wine Flavor and Phenolic Substances.* ACS Symposium, no. 26, American Chemical, 1976, p. 47.

Sols, A.: Selective fermentation and phosphorylation of sugars by Sauterne yeast. *Biochim Biophys Acta, 20:*62, 1956.

Sponholz, W. R. and Dittrich, H. H.: Die Bildung von SO₂-bindenden Gärungs-Nebenprodukten, höheren Alkoholen und Estern bei einigen Reinzuchthefestämmen und bei einingen fur die Weinbereitung wichtigen "wilden" Hefen. *Weinwissensch, 29:*301, 1974.

Stevens, R.: Formation of phenethyl alcohol and tyrosol during fermentation of a synthetic medium lacking amino acids. *Nature, 191:*913, 1961.

Suomalainen, H. and Ronkainen, P.: Mechanism of diacetyl formation. *Nature, 220:*792, 1968.

Tabachnik, J. and Joslyn, M. A.: Formation of esters by yeast. *J Bacteriol, 65:*1, 1953.

Tahara, S., Kurogochi, S., Kudo, M., and Mizutani, J.: Fungal metabolism of sorbic acid. *Agric Biol Chem, 41:*1635, 1977.

Tanner, H.: Der Weinböckser, Entstehung und Beseitigung. *Schweiz Z Obst Weinbau, 105:*252, 1969.

Temperli, A. and Künsch, U.: Die Veränderung der Gehalte an freien Amino-säuren während der Vinifikation. *Qual Plant, 26:*143, 1976.

Thoukis, G., Ueda, M., and Wright, D.: The formation of succinic acid during alcoholic fermentation. *Am J Enol Viticult, 16:*1, 1965.

Thornton, R. J.: Investigation on the genetics of foaming in wine yeasts. *Eur J App Microbiol Biotechnol, 5:*103, 1978.

3-(methylthio) propylamine as a flavor enhancer. Fr. Pat. 1,503,261, 1967.

Troost, G.: Die Technologie des Weines. Stuttgart, Verlag Eugen Ulmer, 1961.

Usseglio-Tomasset, L.: L'acetato d'etile e gli alcooli superiori nei vini. *Riv Viticolt Enolog, 24:*3, 1971.

Van Wyk, C. J.: Malo-lactic fermentation in South African table wines. *Am J Enol Viticult, 27:*181, 1976.

Vetsch, U. and Lüthi, H.: Farbstoffverluste während des biologischen Säureab-baus von Rotweinen. *Mitt Gebiete Lebensm Hyg, 55:*93, 1964.

Wahab, A., Witzke, W., and Cruess, W. V.: Experiments with ester forming yeasts. *Fruit Prod J, 28:*198, 202, 1949.

Wainwright, T.: Hydrogen sulphide production by yeast under conditions of methionine, pantothenate or vitamin B₆ deficiency. *J Gen Microbiol, 61:*107, 1970.

Webb, A. D.: Wine Flavor: volatile aroma compounds of wines. In Schultz, H. W. (Ed.): *Chemistry and Physiology of Flavors.* Westport, Connecticut, Avi, 1967a.

————: Some aroma compounds produced by vinous fermentation. *Biotechnol Bioeng, 9:*305, 1967b.

Webb, A. D. and Ingraham, J. L.: Fusel oil. *Adv Appl Microbiol, 5:*317, 1963.

Webb, A. D. and Kepner, R. E.: Some volatile aroma constituents of *Vitis vinifera var. Muscat of Alexandria. Food Res, 22:*384, 1957.

————: Fusel oil analysis by means of gas liquid partition chromatography. *Am J Enol Viticult, 12:*51, 1961.

Webb, A. D., Kepner, R. E., and Galetto, W. G.: Comparison of the aromas of flor sherry, baked sherry and submerged-culture sherry. *Am J Enol Viticult, 15:*1, 1964.

————: Volatile components of sherry wine. Isolation and identification of N-(2-phenethyl) acetamide and N-isoamyl-acetamide. *Am J Enol Viticult, 17:*1, 1965.

Webb, A. D., Kepner, R. E., and Maggiora, L.: Sherry aroma. VI. Some volatile components of flor sherry of Spanish origin. *Amer J Enol Viticult, 18:*190, 1967.

Webb, A. D. and Muller, C. J.: Volatile aroma components of wines and other fermented beverages. *Adv Appl Microbiol, 15:*75, 1972.

Webb, A. D. and Noble, A. C.: Aroma of sherry wines. *Biotechnol Bioeng, 18:*939, 1976.

Whiting, G. C. and Coggins, R. A.: Organic acid metabolism in cider and perry fermentations. *J Sci Food Agric, 11:*337, 1960.

Williams, P. J. and Strauss, C. R.: The influence of film yeast activity on the aroma volatiles of flor sherries. *J Inst Brew, 84:*148, 1978.

Windholz, M. (Ed.): *The Merck Index, an Encyclopedia of Chemicals and Drugs,* 9th ed. Rahway, New Jersey, Merck Co, 1976.

Würdig, G. and Schlotter, H. A.: Über das Vorkommen SO_2-bildender Hefen in natürlichen Hefegemisch des Traubenmosts. *Dtsch Lebensmittelrundsch, 67:*86, 1971.

Yang, H. Y.: Effect of pH on the activity of *Schizosaccharomyces pombe. J Food Sci, 38:*1156, 1973.

Zimmerman, R.: Über phenolspaltende Hefen. *Naturwissensch, 7:*165, 1958.

Chapter 8

FERMENTED BEVERAGES AND THEIR FLAVOR — II. BEER

They who drink beer, will think beer.

—WASHINGTON IRVING, *The Sketch Book*

INTRODUCTION

MODERN BEER MAKING has led to a change in the production of beer, not only with regard to the quantities of beer brewed in various countries but also in the quality of its taste and aroma; unfortunately these changes are not always for the better. The introduction of the continuous fermentation of beer has so far only aggravated this problem. A better knowledge of the factors involved in the production and enhancement of beer flavor is therefore imperative.

Beer is essentially a fermentation product in which malted grain, usually barley, is subjected to — (1) conversion of starch to fermentable sugars, mostly maltose, by amylolytic enzymes (malted grain has a much stronger liquefying α-amylase enzyme than unmalted grains) and (2) the fermentation of maltose and other fermentable sugars to alcohol. There are several types of beers; the most important ones are ale and lager. Both types of beers are the product of yeast fermentation. Ale is fermented by the yeast *S. cerevisiae,* which rises with the foam during the vigorous fermentation as a thick brown cream. Lager beers are fermented with *S. carlsbergenis* (although Lodder includes it in the species *S. uvarum,* brewery microbiologists still prefer the older name) which accumulate at the bottom of the vessels, hence also bottom fermentation. The top fermentation is carried out at much higher temperatures (15 to 25°C) and leads to a practical exhaustion of fermentable sugars in the wort, while bottom fermentations are run at much lower temperatures (5 to 15°C), hence a much slower fermentation. In the latter case, the fermentation is interrupted while some sugars are still unfermented. The beer is transferred to new vessels at even lower temperatures where the fermentation is led to completion after prolonged storage (lagering). Under these conditions, the typical, more refined taste and aroma of the lager beer is obtained. In addition to water and malt, the brewing wort will also contain hops or hop extracts which contribute to the characteristic bitter taste of beer; it may also contain other unmalted, starchy ingredients from barley or other cereals.

Basically, beers differ from wine not only in the starchy raw materials but also in the fact that it is not dependent on perishable agricultural products, variable with location and season, but rather can be produced practically everywhere under well-defined, controllable conditions. Most beers are now manufactured with a heavy yeast inoculum (pitching) and not by a spontaneous fermentation. This trend in the technology of brewing is very reminiscent of the evolution of bread making. Here a mixed (lactic-alcoholic) fermentation gave rise to a yeast-alcoholic fermentation. The brewing of acid beers (pH = 3.7 and below) is still practiced in many developing countries (Kretschmer, 1978). Further details on the technology of brewing may be found in the classical treatise by DeClerk (1957).

While we are mainly concerned with the contribution of microbes to beer flavor, it is clear that the overall flavor of beer is due to a complexity of factors comprising raw materials, type of malt and the degree of kilning, amount and nature of adjuncts, hops, water, etc. to which we shall occasionally refer. Although many of the aroma and taste compounds created or transformed by microorganisms that we have encountered in the chapter on wine may also contribute to the flavor of beers, no doubt much of the taste (less of aroma) is controlled by the nature of the ingredients.

BREWER'S YEASTS

From the microbiological point of view, *S. cerevisiae* and *S. carlsbergensis* are the most important species in brewing. The major difference between those two yeasts is their fermentation of the triglycoside raffinose, which *S. carlsbergensis* can ferment completely (three-thirds), while *S. cerevisiae* will ferment only in part (one-third) due to the absence of the enzyme melibiase. It is not excluded that these two yeasts are closely related and may have developed as a result of a suitable mutation. It will, however, be difficult to employ these fermentation criteria in the case of hybrids that have become available in the brewing industry. Immunological techniques have been introduced for the distinction between various yeast strains (Campbell, 1972; Haikara and Enari, 1975).

One of the characteristics of yeasts that is of great importance in brewing, more perhaps than in any other fermentation industry, is the phenomenon of flocculation. By this term we mean the ability of certain yeast strains to clump together at a certain stage of their growth curve. While nonflocculating yeast (powdery, *Staubhefe*) will stay in suspension in wort or a suitable buffer, the flocculating yeast *(Bruchhefe)* will sediment to the bottom as can be visualized in a suitable test (Burns, 1937). The aggregation of yeast cells requires certain environmental conditions but seems to be under genetic control (Jansen, 1958; Geilenkotten and Nyns, 1970a; Rainbow, 1970). No doubt, flocculation is a surface phenomenon that can be promoted or abolished by suitable enzymic treatments which expose or eliminate certain functional groups (Geilenkotten and Nyns, 1970b; Nishihara et al., 1977). The ability of flocculating yeast to synthesize higher amounts of the glucomannan complex that constitute the cell wall seems to be in accordance with this concept (MacWilliam and Clapperton, 1969). Flocculent yeast cell walls contain a higher level of phosphate groupings in the outer layer of the mannan-protein envelope. The level of phosphate in the outer layer seems to be sufficient in the case of flocculating yeast strains to enable binding of bivalent cations between adjacent cells. The problem of aggregation and flocculation in biological systems has been recently reviewed by Atkinson and Daoud (1976). Obviously, flocculation decreases the surface area of the yeast population, decreasing the fermentation capacity (less attenuation) and possibly also accelerating some autolytic process (yeasty flavor). Both factors will eventually affect the flavor of the product.

However, the difference in the floc-forming ability of yeasts has often been, and still is, erroneously connected with the phenomenon of top and bottom fermentations. There could be no easier explanation to the rising of the top yeast than the flocculation phenomenon, which engulfs certain gas bubbles that make them rise easily with the foam. The truth is that both fermentations can be carried out with both powdery and flocculating yeasts and that powdery top yeasts also rise to the surface (Burns, 1937). This brings us to the nature of the rising of top yeast. The physical cause of why top yeasts rise to the surface of the wort while bottom yeasts fall to the bottom is rather obscure. It is apparently not a matter of specific gravity of the cells because one would expect all yeasts to rise to the surface if we employ suitable high gravity wort. Others believe that the daughter cells of *S. cerevisiae* remain attached to the mother cells longer, thus forming a light network that makes it easier for gas bubbles to carry them upward. It is possible that the affinity of top yeast cells to gas bubbles is higher in comparison to bottom yeasts. One factor that may not be considered sufficiently in this respect is the difference in the growth rate of these organisms. According to Masschelein et al. (1963), *S. carlsbergensis* has a generation time of 2.2 to 2.4 hours as compared to 1.15 to 1.3 hours of *S. cerevisiae* (at 30°C). *S. cerevisiae* will therefore grow much faster and thus, together with the higher temperatures, will give a higher rate of fermentation in the vessel, which in the end will favor the rise of the top yeast to the surface. Additional studies should clarify this point.

Coming back to the flocculation phenomenon, it is clear that the degree of flocculation of a certain yeast strain employed in brewing will affect the flavor of the product, not only through its attenuation (less alcohol — more residual extract, more "body") but also due to the ease of which yeasts may be separated from the brew (racking). In the case of the secondary fermentation for lager beers, strong flocculation may be a drawback since not enough of the yeast required may be left after the transfer to the new vessel. The choice of the right strain must therefore bring into consideration all the technological aspects of brewing. Gilliland (1951), who classified the yeasts according to their flocculating ability, recommends the use of his Class II yeasts which initially are completely dispersed but form small, loose clumps toward the end of the fermentation.

Attempts to improve the fermentation of wort have been made by hybridization of various strains and the isolation of new species. An interesting organism, *S. diastaticus,* was isolated by Andrews and Gilliland (1952). This yeast differs from other species in its ability to also ferment dextrins and starch to some extent. Fermentation of wort with this organism leads to superattenuation of the beer, i.e. higher alcohol and much lower residual extract values, which are not considered as a technological advantage. An extensive review on the microbiology of brewer's yeasts may be found in several textbooks (Rainbow, 1970; Reed and Peppler, 1973).

FLAVOR IN BEER

The primary contribution of brewer's yeast to the taste of beer is, of course, ethanol. This is another feature in which beer differs from wine since the concentration of ethanol is usually much inferior to that of wine. The concentration of ethanol in beers varies a great deal according to the gravity of the wort employed for the specific beer, generally between 2 and 6 percent (w/w). However, the characteristic flavor of a certain beer is due to a greater extent to other products of fermentation, in addition to Maillard reactions during wort boiling, and hop products (Tressl et al., 1974, 1977). It is generally agreed that higher alcohols (fusel oil), aldehydes, ketones, lower fatty acids, esters etc. determine the taste and aroma of a certain brand. We have already mentioned the importance of the concentration of ethyl alcohol on the threshold values of certain flavor constituents (page 194), which will be modified according to the alcohol content of the specific brand.

Employing conventional methods of analysis, Hartong (1963) suggested an analytical profile of beer aroma composed of the following groups: (a) higher alcohols, esters, aldehydes, and volatile acids, occurring at mg/liter concentrations, yielding a more or less favorable beverage; (b) diacetyl, hydrogen sulfide, and mercaptans appearing at μg/liter quantities that confer an unpleasant aroma to the product. Recent advances in the techniques of flavor analysis (GLC and high pressure liquid chromatography) have greatly enhanced our knowledge of the minor components of beer and their relation to flavor.

Bavisotto and Roch (1959) were probably the first to have studied the beer fermentation employing gas chromatography. Following the formation of fusel oil alcohols, it was found that isoamyl alcohol could be identified very soon after the pitching of the wort, reaching maximum values at the time of the major fermentation period, i.e. within 3 to 5 days, comprising about 60 to 70 percent of the total fusel oil alcohols (Arkima and Sihto, 1963). Pfenninger (1963) analyzed a large number of Swiss beers and found the fusel oil content to vary between 73 and 129 mg/liter. Further, a strong correlation between ethanol production and fusel oil formation could be established, thus confirming the metabolic relationship between these products of yeast activity. There seems to be little change in the fusel oil concentration during the lagering period (Hashimoto and Kuroiwa, 1966). A number of interesting points with regard to the nature of beer fermentation and its effect on the formation of fusel oil were raised by Hough and Stevens (1961), who found that top beers always contained higher levels of fusel oil than bottom beers. These results were confirmed by Suomalainen and Ronkainen (1968) (Table LII).

Considerable work has been done with regard to the effect of temperature on the production of fusel oil by brewer's yeast. As mentioned earlier, the lower concentration of ethyl alcohol in beers results in a greater sensitivity of consumers to fusel oil, hence, the widespread interest in the formation of these alcohols in beers. Fermentations carried out at higher temperatures usually yield beers with higher fusel oil concentrations. Drews et al. (1964) examined a large number of German lager beers and

TABLE LII
FUSEL OIL COMPOSITION OF DIFFERENT BEERS

Type of Beer	Alcohol, mg/liter				
	n-propyl-	*isobutyl-*	*active amyl-*	*isoamyl-*	*phenethyl-*
Bottom-fermented average, 5 types	13	12	14	44	20
Top-fermented average, 5 types	22	38	23	62	41

SOURCE: Adapted from Suomalainen and Ronkainen (1968).

TABLE LIII

EFFECT OF TEMPERATURE ON FUSEL OIL COMPOSITION*

Alcohol	*at 10°C*	*at 25°C*
Propanol	9.5	14.5
Isobutanol	8.1	9.6
2-methyl butanol	15.3	20.7
3-methyl butanol	30.4	47.5
2-phenylethanol	4.6	28.5
	67.9	120.8

SOURCE: Adapted from Engan (1974).

* ppm

found that a change in temperature from 7.5° to 10°C resulted in an increase in the higher alcohol content of the final product from 59 to 77 ppm. Gracheva et al. (1970) could confirm this observation. Using pure culture laboratory fermentations, it was observed that *S. carlsbergensis* strain 11 would yield almost twice as much higher alcohols when the temperature was changed from 2° to 20°C. Above that there was a decrease in fusel oil formation reaching very low levels at 30°C, mostly at the expense of isoamyl alcohol. Engan (1974) extended these studies by including phenethyl alcohol in his analysis (Table LIII).

Clearly, fermentations at higher temperatures result in a sharp rise in the concentration of phenylethanol, with its rosy fragrance that we also encounter in the top-fermented beers. If we consider the threshold concentration of the main fusel oil alcohols in beer to be somewhere between 70 and 100 ppm (Harrison, 1970), and if we consider the possible additive effects between the different fusel oil alcohols, the importance of these alcohols to the overall flavor of beer cannot be neglected.

We have already mentioned the basic pathways for the biosynthesis of higher alcohols by yeasts in another section. Since most authors agree that these pathways also occur in brewer's yeast, we shall not discuss them separately. Although one might expect different yeast strains to produce different yields of higher alcohols, this usually is not the case. An exception was reported by Thorne (1966) in which about 200 ppm of higher alcohols were found with certain strains. Also, the flocculating power of yeast seems to affect little the formation of fusel oil alcohols (Drews et al., 1969). The effect of aeration and stirring on the production of these alcohols in brewing were studied by Maule (1967) and other workers (Lie and Haukeli, 1973). Since increased oxygen availability will stimulate yeast growth, one would expect higher concentrations of fusel alcohols. This happened, in fact, but n-propanol increased more than the other alcohols.

Genetic manipulations of yeast and their possible effect on the formation of fusel alcohols was studied by Ingraham and associates (1961). These workers prepared a number of auxotrophic yeast mutants and studied their fermentation metabolites. Significant differences were found between the yields of the different alcohols. A leucineless mutant could not synthesize isoamyl alcohol, while an isoleucineless strain could not form 2-methyl butanol. A triple auxotroph requiring leucine, isoleucine, and valine would not produce isoamyl, 2-methylbutanol and isobutanol but formed large amounts of n-butanol, which the wild type produced only in trace amounts. These observations may become of greater importance in future studies on the development of better flavor under controlled beer fermentations.

The glycerol content of various beers was investigated by Enebo (1957) who found it to vary between 1500 and 2000 ppm. It is questionable whether glycerol at these concentrations affects the flavor of beers. A number of organic acids comprising acetic, lactic, and traces of other acids were found in beer. As beers are quite acid (pH 3.8 to 4.2), it is clear that these acids contribute to the acid sensation of the product. Since beer fermentations are not run under aseptic conditions, it is not certain to what extent yeast contributes to the acidity of beers through the production of CO_2 and organic acids (Dellweg, 1975) and what role the adventitious bacteria play, at least during the early stages of the fermentation (Eschenbecher and Ellenrieder, 1975). Fatty acids up to C_{18} (mainly caprylic) as well as several keto acids have been described (Suomalainen and Ronkainen, 1968), but we do not know to what extent these compounds, which appear at ppm and less concentrations, affect beer flavor. Recently, it has been shown that the caprylic flavor, a distinctive overall flavor of many commercial beers, is due to the formation of octanoic and decanoic acids during the beer fermentation. This flavor tone is more frequent in lager beers than in ales and has been explained by the fact that lager yeasts (*S. carlsbergensis,* now *S. uvarum*) produce higher amounts of these fatty acids than *S. cerevisiae* (Clapperton and Brown, 1978). Brewer's yeasts may produce 3.6 to 6.4 ppm C_8 and 0.4 to 0.6 ppm C_{10} fatty acids. Considering the threshold values for octanoic and decanoic acids (about 4.5 and 1.5 mg/liter respectively), this "goaty" flavor may be of great significance in beer flavor.

Esters are another important component of the flavor of beers. Ethyl acetate is the most abundant ester followed by smaller amounts of ethyl formate, isoamyl acetate, and other esters (Masschelein et al., 1965). Jenard and Devreux (1964) give a maximum of 27 ppm for bottom beer against 82 ppm for top fermented beers. Since most of the ester formation takes place during the active part of the alcoholic fermentation and almost none during the lagering period, there is little doubt that ester formation is the result of microbial activity. Nordstrom (1965) elucidated the formation of esters by brewer's yeast. Since the formation of ethyl acetate is constant and independent on the concentration of acetic acid in the medium, an endogenous source for the acetate moiety was suggested. CoA, from which acyl-CoA is formed, plays a central role in the mechanism of ester formation. A fruity flavor has been ascribed for beers with high ethyl acetate and isoamyl acetate levels (Gilliland and Harrison, 1966). An undesirable tone of a Finnish beer was traced to a high level of isoamyl acetate. A distinct improvement in the taste of this beer could be established when the yeast strain was changed (Sihto and Arkima, 1963). Yeast strains apparently differ a great deal in their esterogenic activities. Ester formation is highly dependent on temperature and rises up to 25°C. Hence, top-fermented beers have higher ester levels than bottom-fermented beers. The concentration of these esters, including phenethyl acetate, may be above the threshold levels observed in beer (Drews et al., 1969). Thirty ppm and 2 ppm are the commonly accepted levels of ethyl and isoamyl acetate, respectively, in beer (Harrison, 1970).

Acetaldehyde is considered to be a leakage product of the alcoholic fermentation and may attain considerable levels in beer. Maximum values of acetaldehyde have been recorded during the early days of the primary fermentation of lager beer but rarely exceed the threshold concentration in beer in the range of 25 to 50 ppm (Harrison, 1970). However, a "green" off-flavor of beer has been attributed to high acetaldehyde levels. The disappearance of this off-flavor during the latter part of the fermentation and during storage has been correlated with a decrease in acetaldehyde (Sandgren and Enebo, 1961).

Although diacetyl occurs in beer only at very low concentrations, it may be the cause for certain flavor defects. West and coworkers (1952) analyzed a large number of beers and found that diacetyl at concentrations above 0.5 ppm is responsible for the buttery off-flavor of beers. Evidently, the sensitivity to diacetyl in beer is much greater than in the case of wine, where concentrations below 1.0 ppm usually do not impair the product. Although diacetyl is considered a normal by-product of yeast metabolism, the excessive formation of this compound in beer may be connected with a bacterial activity of contaminants such as *Pediococcus, Aerobacter,* etc. (Sandgren and Enebo, 1961). Interestingly, high levels of diacetyl in beer were also found when respiratory deficient mutants of *S. cerevisiae* were employed (Gilliland and Harrison, 1966). Gjertsen et al. (1964) believe that more diacetyl is produced during the fermentation with flocculent yeast than with powdery strains. This opinion was not shared by Makinen and Enari (1969). Diacetyl production is greatly affected by the composition of wort. According to Portno (1966), worts with a low ratio of amino acids to fermentable sugars yield more diacetyl than brews with a high content of amino acids. This was suggested to be due to a feedback inhibition by valine which is related to the biosynthesis of α-acetolactate, the precursor of diacetyl (Fig. 29). Strains of *S. carlsbergensis* are apparently more potent diacetyl producers than *S. cerevisiae*. The possibility that diacetyl production may be a symptom of pantothenate deficiency in the wort rather than valine deficiency has also been suggested (Grinbergs et al., 1968).

A number of methods have been suggested to reduce the diacetyl content of beers in order to reach organoleptically acceptable levels of 0.2 to 0.3 ppm. From the present standpoint,

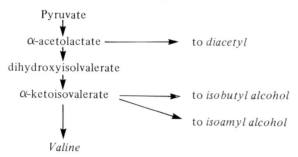

Figure 29. The Biosynthesis of Diacetyl in Beer. Adapted from Ramos-Jeanhomme and Masschelein (1977).

the most interesting method would be the practice of krausening (the addition of an actively fermenting wort to a quiet beer), which promotes a more reductive environment (Lawrence, 1964) and decreases the level of diacetyl. A number of mutant strains devoid of acetolactic synthesis, which produce less than 0.1 ppm vicinal diketones (diacetyl and 2,3 pentadione), have recently been described by Ramos-Jeunehomme and Masschelein (1977).

NEW TECHNOLOGIES

The introduction of new methods for the production of a common beverage like beer is the result of the continuous efforts of fermentation technologists to reduce expenses by saving time, labor, and energy costs. These innovations have to be carefully controlled from the standpoint of flavor, which is the ultimate proof for the success or failure of a new method.

The technology of beer production by the continuous fermentation process poses no particular problems (Bishop, 1970); however, there still remain some uncertainties with regard to the suitability of this technique to the flavor of the product when compared to beers fermented in the conventional batch fermentation. Sandgren and Enebo (1961) found that continuous fermentations produce beers with higher levels of volatile acids, acetaldehyde, and fusel oil alcohols including phenethyl alcohol, bestowing a more alelike flavor to bottom-fermented beer. Since continuous fermentations employ somewhat higher fermentation temperatures and some aeration (Makinen and Enari,

1969), it is not surprising that increased levels of these flavor compounds are formed. The same applies to the somewhat higher levels of diacetyl. Makinen and Enari (1969) claim that when beer is produced by continuous fermentation without aeration, levels of fusel oil alcohols and diacetyl similar to those formed in conventional beer production may be obtained. On the other hand, there was a significant increase in the concentration of esters, particularly isoamyl acetate (up to 11 ppm), when the fermentation was run without aeration. This occurred in spite of the fact that *S. carlsbergensis* produces much smaller amounts of isoamyl acetate than wine yeasts (0.1 ppm vs. 0.32 ppm) when examined in a laboratory sugar fermentation test (Nykänen and Nykänen, 1977). A harsher flavor has been attributed to beers with high ester content. Interestingly, beers produced by the continuous fermentation process contained significantly fewer fatty acids (C_3 through C_{10}) than in batch fermentation. Since the flavor threshold of caprylic acid (caprylic taste) is in the vicinity of 10 ppm, values of 7.2 to 9.1 against 10.6 to 12.2 in the batch fermented beers may be significant from the point of view of taste and aroma. The suppression of ester formation in the presence of long chain, unsaturated fatty acids, as shown by Äyrapää and Lindstrom (1970), would be in accordance with these observations.

This brings us to another innovation in brewing technology and its effect on beer flavor, viz. the fermentation of high gravity wort. In order to increase output and reduce production costs, many breweries have introduced the fermentation of worts of 1.080 instead of the conventional 1.040 gravity, with subsequent dilution of the fermented brew with water. Such beers are rated smoother in taste (Pfisterer and Stewart, 1976). This procedure usually leads to much higher levels of ethyl and isoamyl acetate. The effect of wort gravity on the levels of these esters in beer has been attributed to the deficiency of dissolved oxygen in the more concentrated wort: 5.4 ppm (at 1.080) against 7.1 (at 1.040) at 15°C (Kirsop, 1974). The production of yeast mass during such a fermentation is therefore less than that expected from the higher availability of nutrients in the concentrated wort. According to Anderson and Kirsop (1975), levels of esters may be drastically reduced if suitable oxygenation is carried out in the

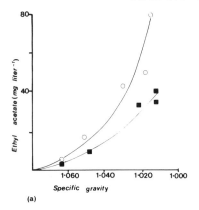
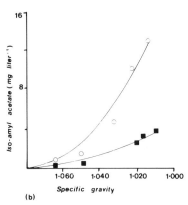

Figure 30. The Effect of Oxygen Supplied at 2, 24, and 44 Hours After Pitching on the Synthesis of: a. Ethyl acetate, b. Isoamyl acetate. Strain N.C.Y.C. 240. Closed circle: pulsed with O_2, open circle: pulsed with N_2. After Anderson and Kirsop (1975).

fermenting wort (30 minutes at 2, 24, and 44 hours). Although considerable variations were observed with various yeast strains, the effect of oxygen on ester formation is quite evident (Fig. 30). These studies should be further extended with regard to the effect of such oxygenation on the biosynthesis of diacetyl and the overall effect on beer flavor.

An interesting new method for the continuous production of beer, employing a reactor in which yeast cells are fixed on a support of PVC and *Kieselgur,* has recently been described (Navarro et al., 1976). Although it is claimed that beers of normal composition were obtained by this procedure, further analytical data on the composition of minor flavor components would be welcome.

SULFUR COMPOUNDS

The role of sulfur compounds in beer flavor and the formation of various sulfur compounds in beer have been reviewed by several authors (Lewis and Wildenradt, 1969; Kuroiwa and Hashimoto, 1970). Analytical detection of these compounds is difficult, and techniques involved in their resolution may result in the appearance of artifacts during the preparation of samples for analysis. Add to this the minute amounts of these compounds in beer and their comparative instability. It is therefore not surprising that conflicting reports on the occurrence and importance of sulfur compounds abound in published literature.

Sulfur-containing volatiles are usually present in beers at very low concentrations (ppb) and constitute only a minor fraction of the total beer sulfur compounds (Kuroiwa and Hashimoto, 1970). They are generally considered to be produced during the active part of the fermentation. These compounds comprise hydrogen sulfide, mercaptans, dimethyl sulfide, dimethyl disulfide, diethyl sulfide, diethyl disulfide, thioformaldehyde, and thioacetone. In addition, the following compounds were recently described —

Methional (3-(methylthio)-propanal)	$CH_3-S-CH_2-CH_2-CHO$
Methionol (3-(methylthio)-1-propanol)	$CH_3-S-CH_2-CH_2-CH_2OH$
Acetic-3-(methylthio)-propyl ester	$CH_2-COOCH_2-CH_2-CH_2-S-CH_3$
3-(methylthio)-propionic acid ethyl ester	$CH_3-S-CH_2-CH_2-COOC_2H_5$
2-methyl-thiophane-3-on	

The biosynthesis of these sulfur compounds from methionine by yeasts has been studied by Schreier et al. (1974, 1976).

Methionol appears in German beers at a concentration of about 1000 μg/liter, but values of over 3000 μg/liter have been detected in the top-fermented wheatbeer (Schreier et al., 1975). Also, acetic-3-(methylthio) propyl ester was higher in the wheatbeer (134 ppb versus 20 to 30 ppb in other types). Considering the fact that methionol has a threshold value in the vicinity of 500 ppb, these compounds cannot be ignored during the evaluation of beer flavor.

It is very difficult to establish the origin of these sulfur compounds in beer, although sulfates are probably only of minor importance (Kuroiwa and Hashimoto, 1970). The biological transformations that lead to the production of some of these compounds have been discussed in earlier chapters (cheese, wine). Clearly, in the case of brewing, the picture is more complicated due to the substantial contribution of precursors (sulfur-containing amino acids) from the plant material and the enzymatic and nonenzymatic reactions that take place during the malting-kilning and mashing processes prior to the fermentation itself. During the initial phase of beer fermentation, there is a sharp rise of H_2S concentration followed by a steady decrease toward the end of the fermentation and during lagering (Kuroiwa and Hashimoto, 1970). However, some authors doubt whether yeasts are at all concerned with the production of organo-sulfur components (Priest et al., 1974). The crucial point is, of course, to what extent do these compounds affect the flavor qualities of beer. Thorne (1966) observed that H_2S at a concentration of 5 ppb had a desirable effect on beer flavor but at 50 ppb imparted a rotten egg odor. Unfortunately, much of the analytical work published on sulfur compounds in beer does not correlate concentrations with a sensory evaluation of the product.

Refined analytical methods have yielded more flavor compounds in beers. Tressl et al. (1973) described a series of lactones with different flavor characteristics in beer. Esters of nicotinic acid, 2-amino acetophenone, as well as various acetamides have also been identified and may be related to yeast metabolism (Tressl et al., 1977). Harding et al. (1977) have recently isolated sixteen basic heterocyclic compounds from English beers. A variety of pyrazines and pyridines (mainly methyl pyrazine and acetylpyridine) have been described. Research on the development of these compounds during brewing and the possible involvement of microorganisms in their formation is eagerly awaited.

BEERS PRODUCED BY SPONTANEOUS FERMENTATIONS

Among the less common beers, we shall mention *lambic*, which is produced in Belgium in the neighborhood of Brussels. This beer is characterized by a sweet-sour taste that evolves during a spontaneous fermentation of a grain mash consisting of malted grains and wheat (Van Oevelen et al., 1976). The slow degradation of dextrins during the later stages of the fermentation as a result of the activities of a mixed flora is claimed to allow the development of the typical lambic flavor. The wort is boiled for several hours and is then run into open shallow trays for cooling. During this period, the wort picks up a variety of organisms from the air that is blown over the wort. After that, the wort is run into large casks that are stored at low temperatures. The primary

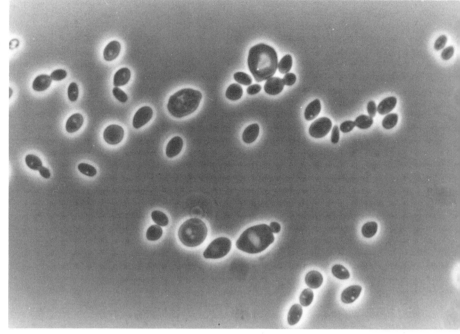

Figure 31a. *Brettanomyces lambicus* (×1250).

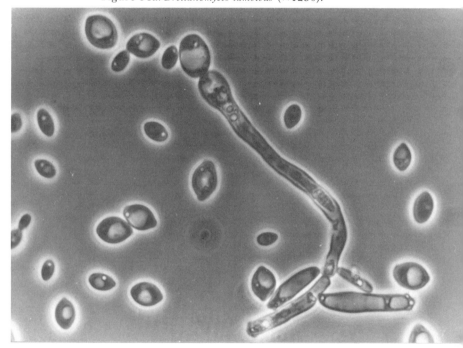

Figure 31b. *Brettanomyces bruxellensis* (×1250). Courtesy of Dr. Van Oevelen, Heveree, Belgium, 1978.

TABLE LIV
THE COMPOSITION OF LAMBIC BEER

Ethanol	4.5-4.6% (w/v)
Propanol	5-9.2 mg/liter
Isobutanol	7-19 mg/liter
Active amyl alcohol	9-16 mg/liter
Isoamyl alcohol	40-58 mg/liter
Phenethyl alcohol	38-64 mg/liter
Acetic acid	530-3944 mg/liter
Lactic acid	492-1344 mg/liter
Ethyl lactate	22-140 mg/liter
Ethyl acetate	12-540 mg/liter
Butanol, isoamyl acetate	traces
pH	3.4-3.9

SOURCE: Adapted from Kretschmer (1977).
NOTE: No doubt the characteristic flavor of lambic beers is due to the high acidities and elevated levels of the corresponding esters.

fermentation starts within a few days and lasts about two months. This is followed by a slower fermentation taking up to two years. Some of the lambic may be bottled after the blending of young and old beer and a third fermentation is taking place in the bottle. After the beer is saturated with CO_2, the new product is marketed under the name of *gueuze*.

The microbiology of the lambic fermentation is a complex one. *Brettanomyces* species, characterized by their ogival cells (Fig. 31), as well as *Saccharomyces* and apiculate yeasts, are accompanied by coliform organisms as well as acetic acid and lactic acid bacteria. During the initial phase, most of the ethanol, as well as higher alcohols, are formed. During the slow fermentation of the second phase, lactic acid is formed, followed by the synthesis of ethyl lactate. Acetic acid is produced within the first few days of the primary fermentation, followed by a gradual formation of ethyl acetate. The composition of lambic beers has recently been described by Kretschmer (1977) (Table LIV).

Rice Wine

Saké is the most widely known fermented beverage of the Far East. Although also known as "rice-*wine*," it is a fermentation product based on cereals but differs from beers in that no malt is used for the conversion of starch to fermentable sugars. Instead, a microbial process involving the fungus *Aspergillus oryzae* is employed. The alcoholic fermentation is carried out by yeast accompanied by some bacteria like *Lactobacillus sake*. Essentially, this is a mixed fermentation and reminds one very much of the fermentation of soy sauce, described earlier (Chapter 5). The gradual liberation of fermentable sugar by the fungus and the alcoholic fermentation by saké yeast (probably a pantothenate-requiring species of *Saccharomyces*) result in high ethanol concentrations which may reach 20 percent (v/v) or more. Analytical studies have shown that most of the fermentation by-products such as fusel oil alcohols, acids, carbonyl and phenolic compounds may also be found in this beverage. Since this fermentation is initiated by the action of mold in the Koji starter, it is not surprising that during the early stages of the fermentation a considerable increase in the concentration of tricarboxylic acids takes place (Owaki, 1967). One might expect an important contribution of phenolic compounds derived from the cereal raw materials to saké flavor; this apparently is the case. Yamamoto et al. (1961) identified ferulic acid, vanillin, and vanillic acid in the final product, while in the Koji stage only ferulic acid could be demonstrated. It was therefore assumed that the transformation of ferulic acid, which constitutes a minor component of the plant material, was carried out by microorganisms involved in the later stages of the saké fermentation. Omori and coworkers (1968) describe the metabolism of ferulic acid by a pure culture of a saké yeast under laboratory conditions. It was found that both p-hydroxybenzoic acid and vanillic acid were formed. When the substrate was changed into vanillin, p-hydroxybenzoic acid, p-hydroxybenzaldehyde, and vanillic acid could be identified. It was concluded that vanillin may be formed as an intermediate in the degradation of ferulic acid, followed by the demethoxylation of vanillic acid.

FERULIC ACID → VANILLIN → VANILLIC ACID → p-HYDROXY BENZALDEHYDE → p-HYDROXY BENZOIC ACID

However, it is known that vanillin is the more important component of saké flavor, contributing to its light and sweet fragrance.

OTHER FLAVOR ASPECTS

Many of the compounds produced by microorganisms during the fermentation of must or wort will also appear in their respective distilled products. Evidently, their concentration depends a great deal on the rectification process so that it is hard to say whether a certain flavor compound in a whisky is due to the specific activity of a certain yeast strain, the availability of nutrients in the specific mash (malt, rye, maize, barley, etc.), or the distillation and aging procedures. For the sake of completion, some analytical data regarding the main flavor components of some commercial whiskies will be reported (Table LV).

A number of phenomena in the beer fermentation have so far received little attention, although they may be related to the activities of yeasts:

1. During the alcoholic fermentation, yeasts produce large amounts of carbon dioxide, which contribute to the formation of foam, which is absolutely necessary in beer consumption. This foam is made of a number of components (carbohydrates, proteins, peptides, glycoproteins) that arise during the fermentation of wort (Archibald et al., 1973; Roberts, 1975). To what extent do yeasts contribute to the formation of foam-building materials? Runkel (1976), in his exhaustive review on foam formation,

TABLE LV

MAIN FLAVOR COMPONENTS OF SOME COMMERCIAL WHISKIES

Whisky	Aliphatic Fusel Oil Alcohols g/liter	β-phenyl Alcohol mg/liter	β-phenylethyl Acetate mg/liter	Ethyl Acetate mg/liter
Scotch	1.02	5.1	2.7	100
Irish	1.77			370
Canadian	0.99	1.8		180
Bourbon	2.17	21.9	2.3	460
Rye	2.50			
Japanese	0.52	2.5	2.2	220

NOTE: Compiled from Suomalainen and Nykänen (1970).

suggested that among other factors that promote foam formation, rapid initial fermentation and high fermentation power of yeasts seem to affect favorably foam formation and stability.

2. A most important component of beer flavor is the bitter fraction derived from hops. These are usually added during the boiling of the mash, and the bitter components together with various terpenes (hop oil) are extracted into the wort. Do yeasts and perhaps other organisms affect the nature or concentration of these bitter principals? Unhopped wort does not yield suitable foam, but yeasts apparently differ in their ability to adsorb the bitter substance, mainly isohumolone (Dixon, 1967). Bottom-fermenting yeasts adsorb much less of the bitter material. The promotion of yeast growth by agitation of the fermenting wort has been shown to decrease the bitterness of beer (Pajunen and Makinen, 1975).

In summarizing the microbial contribution to the taste and

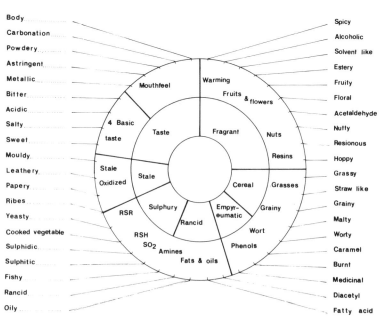

Figure 32. The Flavor Wheel. After Clapperton et al. (1976).

aroma in alcoholic beverages, let us remind the reader that the natural flavor of beer is an extremely complicated balance of the individual effects of different substances, a great number of which are produced by microorganisms during the alcoholic fermentation. One will better appreciate this contribution when glancing at the "Flavor Wheel" (Fig. 32), systematically describing the flavor attributes of beer (Clapperton et al., 1976).

FLAVOR DEFECTS IN BEER

Many of the phenolic compounds we have encountered in the fermentation of plant material, like soy sauce, saké, etc., are indispensable when present in a beverage in limited concentrations but will cause off-flavors at high concentrations.

A deviation from the standard flavor of a certain beer type may occur because of a number of reasons. We shall briefly discuss those defects that are known to be caused by microbial activity. Among these, the phenolic off-flavor, often described as medicinal or pharmacylike, is the most important defect (Dadić et al., 1971). Considering the very low flavor threshold of phenolic compounds (ppb), it is not surprising that such off-flavors may lead to a rejection of such products. According to West et al. (1965), normal beers contain 9.6 to 16.9 ppb phenol and 3.0 to 11.8 ppb chlorophenol, while beers with a perceptible phenolic taste contained 19.0 to 71.2 ppb of phenol and 4.0 to 38.6 ppb of chlorophenols. Similar levels were reported by Wackerbauer et al. (1971). There may be different sources for the phenolic compounds in beer. Contamination from chemical sources (disinfectants) or decaying material, poor water supplies, improper handling of tanks, or can lining will not be discussed here. Phenolic compounds (steam volatile) may also appear at elevated, undesirable levels because of microbial activity on certain components of the wort. Certain organic acids found in grain mashes (ferulic acid, p-coumaric acid), as well as vanillin derived from lignin of the plant material, may be transformed by certain yeast strains and other organisms to form products that possess strong medicinal flavor.

Steinke and Paulson (1964) suggested that ferulic acid yields 4-vinyl guaiacol, p-coumaric acid yields 4-vinyl phenol, while vanillin is transformed to 4-methyl guaiacol. 4-vinyl phenol and

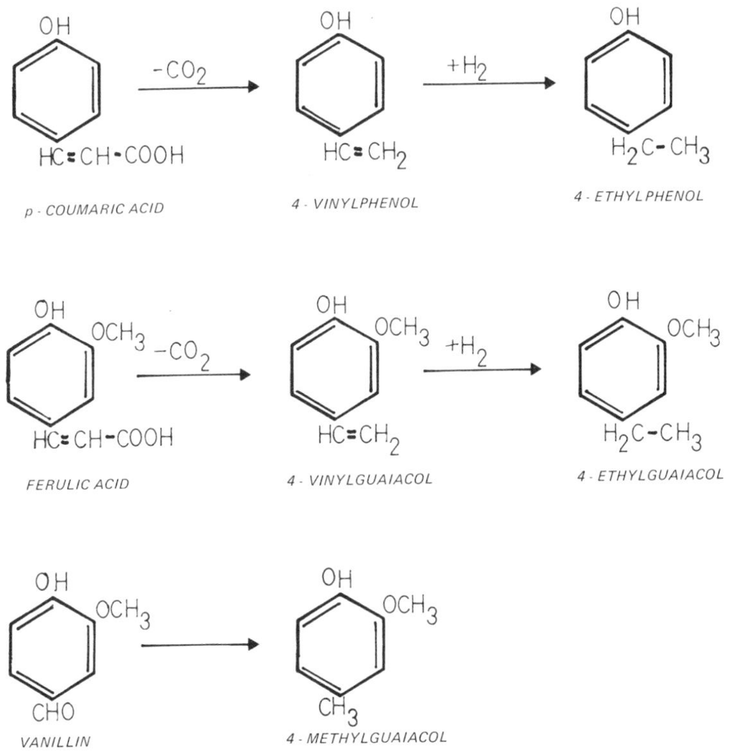

Figure 33. Transformation of Precursors to Steam Volatile Phenols.

4-vinyl guaiacol may be further reduced to 4-ethylphenol and 4-ethyl guaiacol, respectively (Fig. 33). Volatile phenols may also be formed from aromatic amino acids (phenylalanine, tyrosine) through the action of certain bacteria as suggested by Lawrence (1964). According to this author, several gram-negative and gram-positive bacteria partake in the transformation of tyrosine into p-coumaric acid (p-hydroxy phenylacrylic acid), which in turn yields volatile phenols.

The transformation of tyrosine into p-hydroxy phenylacetic acid and to cresol has been ascribed to the activities of certain anaerobes:

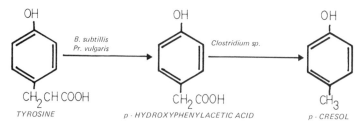

Experimental evidence for the involvement of microbes in the formation of volatile phenolic compounds in beers has been put forward by Riemann and Scheible (1969). These workers carried out a series of pure culture fermentations in wort inoculated with a variety of microorganisms isolated from breweries. Abnormally high phenol contents were found in beer produced from wort infected with yeasts other than brewer's or Termobacteria. The latter is a somewhat unfortunate term used in many industries to designate small, gram-negative bacteria (but has nothing to do with temperature), including species of the coliform group; *Pseudomonas, Zymomonas, Flavobacterium,* and *Achromobacter* (Table LVI).

It is clear that microbial infections by yeasts that are not related to the brewing organism (wild yeast), as well as certain bacteria, affect the taste of beer through the production of certain phenolic compounds, the exact identity of which remains to be clarified. The ability of enterobacterial organisms to produce steam volatile phenolic compounds when grown in wort by decarboxylation of ferulic acid or p-coumaric acid has been examined by Lindsay and Priest (1975). Certain strains of the *Hafnia* group (especially *H. alvei, H. protea*) seem to be more active in this respect. These, as well as other gram-negative bacteria, are also involved in the production of organosulfur compounds in the wort, particularly dimethyl sulfide (Priest et al., 1974).

A phenolic-herbal flavor defect has been described in a two-staged, stirred continuous beer fermentation (Maule and Thomas, 1973). This was found to be due to the overgrowth of certain yeast strains lethal ("Killer") to brewer's yeasts.

Another off-flavor found in stale beer, also referred to as

TABLE LVI
FLAVOR CHARACTERISTICS OF WORT FERMENTED WITH VARIOUS CONTAMINANTS

Organism	Yeast W548	S. ellipsoideus	S. pastorianus	Termobacteria*	Pichia	Torulopsis
Odor	Pure	Musty	Smoky, unpleasant	phenolic (celery)	spicy	meady
Taste	pure	smoky, harsh	harsh	sweet, sour	spicy	spicy
pH	4.33	4.04	4.20	4.47	5.12	4.58
phenol† (µg/liter)	41	103	130	142	46	59

SOURCE: Adapted from Riemann and Scheible (1969).
* Not further identified.
† Approximate composition of phenolic compounds.

cardboard off-flavor, has been ascribed to the appearance of certain aldehydes in beers. Among these, trans 2-hexenal, trans 2-nonenal, and trans 2,4-nonadienal seem to be the most important. According to Dominguez and Canales (1974), these aldehydes are degradation products of wort lipids and arise through a series of oxydation reactions starting from linoleic and linolenic acids. It is suspected that yeast during beer fermentation may provide the necessary enzymes (lipases, lipoxygenases, and hydroxyperoxydases) for the production of these unsaturated aldehydes. However, the involvement of yeast enzymes in these reactions has not yet been fully documented.

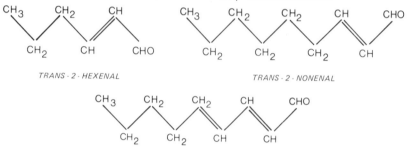

TRANS - 2 - HEXENAL   TRANS - 2 - NONENAL

TRANS - 2, TRANS 4 - NONADIENAL

A phenomenon related to beer fermentation has been recently ascribed to the activity of certain fungi. In a number of breweries the "gushing" of beer after opening a beer bottle has been observed. This may lead to a loss of over one-half of the contents of the bottle. Many authors relate this phenomenon to the contamination of certain fungi (*Aspergillus fumigatus* and *As. amstelodami*) during the malting process (Gyllang et al., 1977). Japanese workers ascribed this phenomenon to the activity of certain fungi (*Nigrospora sp.*) derived from barley grains (Kitabatake and Amaha, 1976). Although little is known about the physicochemical processes that lead to gushing of beer, it is assumed that these changes are brought about by the metabolic activities of fungi. A beer-gushing inducing protein (NGF) was isolated from a culture of *Nigrospora* and was found to be characterized by a relatively high content of hydrophobic amino acid residues (Kitabatake et al., 1977).

Another not very frequent off-flavor in fermented beverages should be mentioned here. Acrolein (CH_2=CHCHO) and acrolein derivatives are sometimes formed during fermentation. Earlier authors indicated that various organisms, e.g. *Cl. welchii, E. freundii, B. amaracrylus (B. polymyxa),* were responsible for the formation of acrolein. Although the mode of acrolein formation by bacteria is not completely understood, most probably it originates from glycerol, another by-product of the alcoholic fermentation. Since acrolein production is a feature of bacterial metabolism, contamination of beverages may be considered the main source. Wine, beer, brandy, and grain fermented distillates may be susceptible to this defect owing to the frequent contamination by lactic organisms (Serjak et al., 1954). When acrolein appears in grain alcohol fermentation, it is called "pepper." The term pepper probably originates from the pronounced irritant and lachrymatory properties of acrolein, especially when it is distilled from affected beers. In commercial distilleries, it occurs sporadically. During the pepper period, alcohol distillates range in their acrolein content from trace to 100 ppm or more. As little as 10 ppm in whisky are readily detectable organoleptically. Such products must be redistilled before marketing. Why is the pepper defect not more frequent? According to Serjak et al. (1954), under normal operating conditions the bacterial contamination reaches its maximum before fermentable sugars are exhausted; under these conditions nonacrolein strains predominate. If, however, lactobacilli develop at subnormal rates (low contamination or other reasons), growth may intensify after carbohydrates are depleted; the glycerol is utilized as a carbon source with the formation of acrolein.

In wines and brandies, a bitterness sometimes appears that has also been related to acrolein. Although the latter compound is not bitter per se, the combination of acrolein with polyphenol compounds such as tannins found in grapes or barrel extracts may lead to this sensation. High tannin wines were found to yield marked bitterness upon the addition of acrolein, whereas low tannin wines were not affected.

Rum suffers occasionally from a similar off-flavor. In this case, two addition products of ethanol and acrolein (2-ethoxy propanol and 1,1,3-triethoxy propane) were identified (Dubois et al., 1973).

BIBLIOGRAPHY

Anderson, R. G. and Kirsop, B. H.: Oxygen as a regulator of ester accumulation during the fermentation of wort of high specific gravity. *J Inst Brew, 81:*111, 1975.

Andrews, J. and Gilliland, R. B.: Super-attenuation of beer. *J Inst Brew, 58:*189, 1952.

Archibald, H. W., Weiner, J. P., and Taylor, L.: Observation of factors affecting beer foam characteristics. *Proc Europ Brew Conv*, Salzburg, 1975, p. 349.

Arkima, V. and Sihto, E.: Über die Gaschromatographie einiger im Verlaufe der Biergärung entstehende Aromakomponenten. *Proc Europ Brew Conv*, Brussels, 1963, p. 268.

Atkinson, B. and Daoud, I. S.: Microbioal flocs and flocculation in fermentation process engineering. *Adv Biochem Eng, 4:*42, 1976.

Äyrapää, T. and Lindstrom, I.: Influence of long-chain fatty acids on the formation of esters by brewer's yeast. *Proc Europ Brew Conv*, Salzburg, 1973, p. 261.

Bavisotto, V. S. and Roch, L. A.: Gas chromatography of volatiles in beer during its brewing, fermentation and storage. *Proc Am Soc Brew Chem*, 1959, p. 63.

Bishop, L. R.: A system of continuous fermentation. *J Inst Brew, 76:*172, 1970.

Burns, J. A.: Yeast flocculation. *J Inst Brew, 43:*31, 1937.

Campbell, I.: Simplified identification of yeasts by a serological technique. *J Inst Brew, 78:*225, 1972.

Clapperton, J. F. and Brown, D. G.: Caprylic flavour as a feature of beer flavour. *J Inst Brew, 84:*90, 1978.

Clapperton, J. F., Dagliesh, C. E., and Meilgaard, N. C.: Progress towards an international system of beer flavor terminology. *J Inst Brew, 82:*7, 1976.

Dadić, M., Van Gheluwe, J. E. A., and Valyi, Z.: Phenolic taste in beer. *Wallerstein Lab Comm, 34:*5, 1971.

De Clerk, J.: *A Textbook of Brewing.* London, Chapman & Hall, Ltd, 1957.

Dellweg, H.: On the occurrence of α-keto acids and organic acids in beer fermentation. *Monatsschr Brauerei, 28:*172, 1975.

Dixon, I. J.: Hop substances and yeast behavior. *J Inst Brew, 73:*488, 1967.

Dominguez, X. A. and Canales, A. M.: Oxidation of beer. *Brewers Dig, 49:*40, 1974.

Drews, B., Bärwald, G., and Niefiend, H. J.: Some metabolic products of beer yeast and their significance in fermentation technology. *Techn Quart MBAA, 6:*193, 1969.

Drews, B., Specht, H., and Bärwald, G.: Higher aliphtic alcohols in beer and some factors which quantitatively influence their formation. *Monatsschr Brauerei, 17:*101, 1964.

Drews, B., Specht, H., and Kühl, E. D.: Sulfur compounds in beer. *Monatsschr Brauerei, 13:*239, 1966.

Dubois, P., Parfait, A., and Dekimpe, J.: Presence de dérivés de l'acroleine dans un rhum a goût anormal. *Ann Technol Agric, 22:*131, 1973.

Enebo, L.: On some volatile minor compounds in beer. *Proc Europ Brew Conv,* Copenhagen, 1957, p. 370.

Engan, S.: Higher alcohols in beer. *Brewers Dig, 49:*52, 1974.

Eschenbecher, F. and Ellenieder, M.: Eine Artenanalyse der Würzebakterien. *Proc. Europ. Brew. Conv.,* Nice, 1975, p. 497.

Geilenkotten, I. and Nyns, E. J.: Biochemie de la floculence, une revue de la littérature. *Bull Assoc Roy Anc Etud Brass, 66:*19, 1970a.

———: Induction de la floculence chez *Saccharomyces cerevisiae* par le suc d'Helix pomatia. *Bull Assoc Anc Etud Brass, 66:*59, 1970b.

Gilliland, R. B.: The flocculation characteristics of brewing yeast during fermentation. *Proc. Eur. Brew. Conv.,* Brighton, 1951, p. 35.

Gilliland, R. B. and Harrison, G. A. F.: Flavor in beer. *J Appl Bacteriol, 29:*244, 1966.

Gjertsen, P., Undstrup, S., and Trolle, B.: A rapid and sensitive method for the determination of diacetyl in beer. *Monatschr Brauerei, 17:*232, 1964.

Gracheva, I. M., Veselov, I. Y., Gavrilova, N. N., and Kovalevich, L. S.: Influence of the temperature of fermentation on the formation of higher alcohols by yeast. *Mikrobiologiya, 39:*322, 1970.

Grinbergs, M., Clarke, B. J., and Hildebrand, R. P.: Secondary flavor notes of beer. *Inst Brewing Proc,* Australia, 1968, p. 23.

Gyllang, H., Satmark, L., and Martinson, E.: The influence of some fungi on malt quality. *Proc Europ Brew Conv,* Amsterdam, 1977, p. 245.

Haikara, A. and Enari, T. M.: The detection of wild yeast contaminants by the immunofluorescence technique. *Proc Europ Brew Conv,* Nice, 1975, p. 363.

Harding, R. J., Nursten, H. E., and Wren, J. J.: Basic compounds contributing to beer flavor. *J Sci Food Agric, 28:*225, 1977.

Harrison, G. A. F.: The flavor of beer, a review. *J Inst Brew, 76:*486, 1970.

Hartong, B. D.: An analytical profile of beer aroma. *Proc Europ Brew Conv,* Brussels, 1963, p. 528.

Hashimoto, N. and Kuroiwa, Y.: Gas chromatographic studies on volatile alcohols and esters of beer. *J Inst Brew, 72:*151, 1966.

Hough, J. S. and Stevens, R.: Beer flavor. IV. Factors affecting the production of fusel oil. *J Inst Brew, 67:*488, 1961.

Jansen, H. E.: Flocculation of yeasts. In Cook, A. H. (Ed.): *The Chemistry and Biology of Yeasts.* New York, Acad Pr, 1958.

Jenard, H. and Devreux, A.: Étude des constituents volatiles de la biere. *Echo Brass, 20:*1119, 1964.

Kirsop, B. H.: Oxygen in brewery fermentation. *J Inst Brew, 80:*252, 1974.

Kitabatake, K. and Amaha, M.: Properties of the two gushing inducing factors of *Nigrospora sp.* No. 207, produced on barley grains. *Bull Brew Soc, 22:*9, 1976.

———: Effect of chemical modifications on the gushing-inducing activity of a hydrophobic protein produced by a *Nigrospora sp. Agric Biol Chem, 41:*1011, 1977.

Kretschmer, K. F.: Das mikrobiologische Phänomen der Aromabildung bei der Selbstgärung des belgischen Lambic-bieres. *Monatsschr Brauerei,* 30:514, 1977.

———: Alte Brauweisen als Förderer menschlicher Gesundheit. *Ulmer Braumeister,* 15:9, 1978.

Kuroiwa, Y. and Hashimoto, N.: Sulfur compounds responsible for beer flavor. *Brewers Dig, 45:*48, 1970.

Lawrence, W. C.: Volatile compounds affecting beer flavor. *Wallerstein Lab Comm, 27:*123, 1964.

Lewis, M. J. and Wildenradt, H. L.: Observations on sulfur in brewing. *Brewers Dig, 44:*88, 1969.

Lie, S. and Haukeli, A. D.: Production of volatiles during yeast fermentation. *Proc Europ Brew Conv,* Salzburg, 1973, p. 285.

Lindsay, R. F. and Priest, F. G.: Decarboxylation of substituted cinnamic acids by enterobacteria; the influence on beer flavor. *J Appl Bacteriol, 39:*181, 1975.

Lodder, J.: *The Yeasts: A Taxonomic Study,* 2d ed. Amsterdam, North Holland Pub, 1970.

MacWilliam, I. C. and Clapperton, J. F.: Dynamic aspects of nitrogen metabolism in yeast. *Proc Eur Brew Conv,* Interlaken, 1969, p. 271.

Makinen, V. and Enari, T. M.: Flavor compounds in continuous fermentation. *Brewers Dig, 44:*72, 1969.

Masschelein, A., Jenard, H., Jeunhomme-Ramos, C., and Devreux, A.: Métabolisme fermentaire et synthése des volatils au cours de la fermentation. *Proc Eur Brew Conv,* Stockholm, 1965, p. 209.

Masschelein, C. A., Jeunhomme-Ramos, C., Jenard, H., Haboucha, J., and Devreux, A.: Biochemical and biophysical characterization of various strains of brewers' yeast. *Proc Europ Brew Conv,* Brussels, 1963, p. 381.

Maule, A. P. and Thomas, P. D.: Strains of yeast lethal to brewery yeasts. *J Inst Brew, 79:*137, 1973.

Maule, D. R.: Rapid gas chromatographic examination of beer flavor. *J Inst Brew, 73:*351, 1967.

Navarro, J. M., Durand, G., Duteurtre, B., Moll, M., and Corrieu, G.: Mise au point d'un procédé continu de préparation de boissous fermentées. *Ind Alim Agric, 93:*695, 1976.

Nishihara, H., Toraya, T., and Fukui, S.: Effect of chemical modification of cell surface components of a Brewer's yeast on the floc-forming ability. *Arch Microbiol, 115:*19, 1977.

Nordstrom, K.: Possible control of volatile ester formation in brewing. *Proc Europ Brew Conv,* Stockholm, 1965, p. 195.

Nykänen, L. and Nykänen, I.: Production of esters by different yeast strains in sugar fermentations. *J Inst Brew, 83:*30, 1977.

Omori, T., Yamamoto, A., and Yashi, H.: Studies on the flavor in saké. *Agric Biol Chem, 32:*539, 1968.

Owaki, K.: Components of saké. IV. Carbonyl compounds. *Nippon Jozo Kyoki Zasshi, 62:*1097, 1967.

Pajunen, E. and Makinen, V.: The effect of pH and agitation on beer fermentation. *Proc Europ Brew Conv*, Nice, 1975, p. 525.

Pfenninger, H. B.: Zur Fuselölanalyse in Bieren. *Proc Europ Brew Conv*, Brussels, 1963, p. 257.

Pfisterer, E. and Stewart, G. G.: High gravity brewing. *Brewers Dig*, *51*:34, 1976.

Portno, A. D.: Some factors affecting the concentration of diacetyl in beer. *J Inst Brew*, *72*:193, 1966.

Priest, F. G., Cowbourne, M. A., and Hough, J. S.: Wort enterobacteria, a review. *J Inst Brew*, *80*:342, 1974.

Rainbow, C.: Brewer's yeasts. In Rose, A. H. and Harrison, J. S. (Eds.): *The Yeasts*, vol. 3, London and New York, Acad Pr, 1970.

Ramos-Jeunehomme, C. and Masschelein, C. A.: Controle genetique de la formation des dicetones vinicales chez *Saccharomyces cerevisiae*. *Proc Europ Brew Conv*, Amsterdam, 1977, p. 267.

Reed, G. R. and Peppler, H. J.: *Yeast Technology*. Westport, Avi, 1973.

Riemann, J., and Scheible, E.: Der Einfluss von Infektionen der Würze bzw des Bieres auf den Geruch und Geschmack des Bieres. *Brauwelt 109*:1074, 1969.

Roberts, R. T.: Glycoproteins and beer foam. *Proc Europ Brew Conv*, Nice, 1975, p. 453.

Runkel, U. D.: Der Einfluss von Rohstoffen und Verfahrensbedingungen auf den Schaum. *Monatsschr Brauerei*, *29*:248, 1976.

Sandgren, E. and Enebo, L.: Biochemical aspects of continuous alcoholic fermentation. *Wallerstein Lab Comm*, *24*:269, 1961.

Schreier, P., Drawert, F., and Junker, A.: Flüchtige Schwefelverbindungen des Bieraromas. *Brauwiss*, *27*:205, 1974.

————: Gas chromatographische Bestimmung der Inhaltsstoffe von Gärungsgetränken. *Brauwiss*, *28*:73, 1975.

Schreier, P., Drawert, F., Junker, A., Barton, H. and Leupold, G.: Über die Biosynthese von Aromastoffen durch Mikroorganismen. II. Bildung von Schwefelverbindungen aus Methionin durch *Saccharomyces cerevisiae*. *Z Lebensm Unters Forsch*, *162*:279, 1976.

Serjak, W. C., Day, W. H., Van Lanen, J. M., Boruff, C. S.: Acrolein production by bacteria found in distillery grain mashes. *Appl Microbiol*, *2*:14, 1954.

Sihto, E. and Arkima, V.: Proportions of some fusel oil components in beer and their effect on aroma. *J Inst Brew*, *69*:20, 1963.

Steinke, R. D. and Paulson, M. C.: The production of steam-volatile phenols during the cooking and alcoholic fermentation of grain. *J Agric Food Chem*, *12*:381, 1964.

Suomalainen, H. and Nykänen, L.: Composition of whisky flavor. *Proc Biochem*, *5*:1, 1970.

Suomalainen, H. and Ronkainen, P.: Aroma components and their formation in beer. *Techn Quart MBAA*, *5*:119, 1968.

Thorne, R. S. W.: The contribution of yeast to beer flavor. *Techn Quart MBAA*, *3*:160, 1966.

Tressl, R., Kossa, T., and Renner, R.: GLC-MS Untersuchungen flüchtiger Inhaltstoffe von Hopfen, Würze und Bier und deren Genese. *Proc Europ Brew Conv*, Salzburg, 1973, p. 737.

————: Über die Bildung von Aromastoffen in Malz, Würze und Bier. *Monatsschr Brauerei*, *27*:98, 1974.

Tressl, R., Renner, R., Kossa, T., and Kopler, H.: GLC-MS Untersuchungen flüchtiger Inhaltstoffe von Hopfen, Würze und Bier und deren Genese. *Proc Europ Brew Conv*, Amsterdam, 1977, p. 693.

Van Oevelen, D., de l'Escaille, F., and Verachtert, H.: Synthesis of aroma components during the spontaneous fermentation of lambic and gueuze. *J Inst Brew, 82:*322, 1976.

Wackerbauer, K., Kossa, T., and Tressl, R.: Die Bildung von Phenolen durch Hefen. *Proc Europ Brew Conv*, Amsterdam, 1977, p. 495.

West, D. B., Lautenbach, A. F., and Becker, K.: Biacetyl in beer. *Proc Am Soc Brew Chem*, Toronto, 1952, p. 81.

West, D. B., Lautenbach, A. F., and Brumsted, D. D.: Phenolic characteristics in brewing. *Wallerstein Lab Comm, 28:*90, 1965.

White, F. H. and Portno, A. D.: Continuous fermentation by immobilized brewers yeast. *J Inst Brew, 84:*228, 1978.

Yamamoto, A., Sasaki, K., and Saruno, R.: Flavors of sake. IV. Separation and identification of ferulic acid, vanillic acid and vanillin. *Nippon Noegeikagaku Kaishi, 35:*715, 1961.

Chapter 9

PRODUCTION OF FLAVOR AND FLAVOR-ENHANCING COMPOUNDS BY MICROORGANISMS

*Not presume to dictate, but broiled fowl
and mushrooms — capital thing!*

—CHARLES DICKENS, *Pickwick Papers*

GENERAL CONSIDERATIONS

MICROORGANISMS play a very important role in the commercial production of flavor compounds, the best known example being citric acid. This compound has become an indispensible flavor ingredient of many foods and beverages and is produced today in bulk quantities that almost make them part of the staple food products. Although citric acid is a typical fermentation product, employing selected strains of *Aspergillus niger,* it will not be considered here further since many technical details of the citric acid fermentation are available in textbooks (Underkoffler and Hickey, 1954; Peppler, 1961). We shall devote this chapter rather to the production of flavor-enhancing compounds that are either produced by microorganisms in minute quantities or are useful in the process of flavor enhancement in very small concentrations.

MUSHROOMS AND FLAVOR

Mushrooms are different from the microbes we have mentioned so far in being fungi that at a certain stage of their life cycle form "conspicuous, fleshy sporofores at macroscopic dimensions" (Worgan, 1968), known to every lover of nature and to many connoisseurs of good food. In spite of the many thousands of mushrooms that can be found in various localities, the number of these known to the wide public is rather limited to those few that are produced on a commercial scale. These in-

256

clude the champignon *Agaricus campestris (bisporus)* of the Western world, cultivated on horse manure; the shiitake (*Lentinus* or *Cortinellus edodes*) of Japan, cultivated on wooden logs; and the Padi straw mushroom *(Volvariella volvacea)* of Southeast Asia and Africa, cultivated on straw (Singer, 1961; Gray, 1972).

Botanically speaking, mushrooms may be of the order Basidiomycetes, to which most of the mushrooms belong, or Ascomycetes, the best known of which are the morels (*Morchella* spp.) and the underground delicious truffles (*Tuber* spp.).

Fresh mushrooms are proteinaceous foods (30 to 50% on dry weight) with an outstanding flavor. This is even stronger when mushrooms are dried and used as powerful flavor enhancers or spices. The increased demand for mushrooms for both domestic use and industrial food formulations have put the price upward, which induced many microbiologists to attempt the production of mushroom flavor by submerged fermentation of such mycelium. Media compositions and fermentation conditions for a wide range of mushrooms have been worked out (Worgan, 1968). The recovery of mushroom mycelium in the form of powder or pellets has also been studied (Robinson and Davidson, 1959).

There would be, no doubt, a large commercial outlet for a dried preparation of mycelial mushroom material that retained the full mushroom flavor. To this end, the fermentation of mushroom mycelia with distinct mushroom flavor is a prerequisite. Although several claims have been made that mycelium with typical flavor have been obtained in submerged culture, it is doubtful if a process for commercial exploitation has been developed that permits the consistent production of mycelia with the flavor complexity and fullness of fleshy mushrooms.

Early literature disclosed conflicting data on the suitability of various species and genera to the submerged cultivation of mushrooms. Some claim more success with morels, others prefer the *Agaricus* species. Sugihara and Humfeld (1954) grew twenty different strains in submerged culture and submitted the harvested mycelium to a taste panel. Descriptions of nutlike, cheeselike, and pleasant were presented, but none were identical with the cultivated mushroom flavor. Only *Lepiota rhacodes* and

A. campestris produced any flavor at all. Eddy (1958) concluded that the production of a full mushroom flavor by submerged culture mycelium is not possible.

Several improvements in the production of mushroom mycelium with typical flavor have been suggested. The incorporation of edible oils or lecithin seem to intensify mushroom flavor (*Morchella esculenta,* Szuecs, 1954). Sugimori et al. (1971) found that submerged mycelium of *Lentinus edodes* had a better mushroom flavor when grown on ethanol than when cultivated on glucose. Gilbert (1960) studied the effect of aeration on the formation of pellets or filamentous growth of *Morchella* species. A good correlation between pellet formation and flavor was found with *Mo. crassipes.* These observations were later confirmed by Lichtfield (1967) who worked with *M. hortensis, M. esculenta,* and *M. crassipes.*

The last decade has evidenced considerable progress in the production of flavor rich mushroom mycelium as a result of both improved fermentation technology and the advent of more refined chemical analysis of mushroom flavor ingredients. Before going into more details about the fermentation of mushroom mycelium, a brief summary of the chemical composition of mushroom flavor is appropriate.

The characterization of the flavor components of fresh and dried mushrooms has been attempted by several workers. Mushroom flavor is probably composed of a number of non-volatile compounds like glutamic acid and 5'-guanylic acid (5'-GMP). These flavor ingredients, better known as flavor enhancers, have a wider importance in the chemistry of flavor and will be discussed at length in a later section. The more characteristic flavor of mushrooms may be attributed to a series of volatile compounds, which probably has a basic pattern common to most mushrooms, with certain fluctuations in concentrations and some specific compounds characteristic to specific varieties or strains.

Finnish chemists have devoted considerable efforts to the analysis of the chemical entities that constitute the flavor of mushrooms. According to Pyysalo (1976) and Pyysalo and Suhiko (1976), the flavor backbone of mushrooms is made of a variety of C_8 compounds, mainly 1-octen-3-ol, which has a typical mushroom odor, while 1-octene-3-one seems to be important in producing the more specific aroma of wild mushrooms. Lower homologues (1-heptene-3-ol etc.) are much less odorous (Ney and Freytag, 1978). Also, other C_8 and C_9 compounds may take part in the formation of this aroma (Table LVII). The occurrence of 1-octen-3-ol in various mushrooms has been described by Dijkstra (1976, Table LVIII).

$$\overset{\displaystyle H}{CH_3(CH_2)_4-\underset{\displaystyle OH}{C}-CH=CH_2}$$

1-OCTENE-3-OL

Considering the fact that mushrooms contain only about 5 to 15 ppm of volatiles, the contribution of these compounds to the flavor of mushrooms must be related to their very low flavor thresholds. According to Varoquaux et al. (1977), linoleic acid serves as precursor of 1-octene-3-ol in *A. bisporus*.

More compounds have been enumerated in various publications on mushroom flavor. Thomas (1973) describes close to seventy components of the flavor of dried *Boletus edulis*, including pyrazines and 2-formylpyrroles:

2 - FORMYL - 5 - (2 - PHENYLETHYL) PYRROLE   *1 - (PHENYLETHYL) - 2 - FORMYLPYRROLE*

In the shiitake mushroom, *Ln. edodes*, a sulfur compound was described by Morita and Kobayashi (1966) as lenthionine:

TABLE LVII

VOLATILES OF FRESH MUSHROOMS, THRESHOLD VALUES, SPECIFIC ODOR CHARACTERISTIC, AND RELATIVE
DISTRIBUTION OF COMPONENTS

Compound	Threshold Value (ppm)	Odor Characteristic	Relative Distribution in Various Species	Highest Concentration Found in
1-octen-3-ol	0.01	mushroomlike	30-90	*Lactarius torminosus*
1-octen-3-one	0.004	like boiled mushrooms	0.02-8	*Boletus edulis*
trans-2-octen-1-ol	0.040	somewhat medicinal, oily	0.02-24	*Cantharellus cibarius*
trans-2-octen	0.003	sweet, somewhat phenolic	0.1-3.0	*Lactarius rufus*
3-octanol	0.018	like cod liver oil	0.02-3.9	*Gyromitra esculenta*
3-octanone	0.050	fruity, musty	0.2-4.0	*Agaricus bisporus*
octanol	0.48	sweet, soapy	0.1-0.9	*Cantharellus cibarius*
1-octene-3-yl acetate	0.09	mushroomlike, somewhat soapy	0.03-0.05	various
1-octene-3-yl propionate	0.022	fruity, mushroomlike	0.1-0.5	*Boletus edulis*
nonanol	0.09	sweet, somewhat soapy	0.05-1	*Gyromitra esculenta*

SOURCE: Adapted from Pyysalo and Suhiko (1976).

which is claimed to be the major aroma bearing substance of this mushroom. The cultivated champignon, *A. bisporus,* differs from wild mushrooms in its high content of benzyl alcohol and benzaldehyde but low concentrations of the C$_8$ components (Pyysalo, 1976). The false morel *(Gyromitra esculenta)* must be dried or cooked before use because of toxic N-methyl-N-formylhydrazones (Pyysalo and Niskanen, 1977). A flavorful lactone, 2-methyl-2-pentene-4-olide was described in *Coprinus comatus* Dijkstra and Wiken, 1976):

TABLE LVIII

1-OCTEN-3-OL IN EXTRACTS OF FRUIT BODIES OF EDIBLE MUSHROOMS

Species	*1-octen-3-OL* $\mu l/l$
Agaricus bisporus, fresh	3.3
Agaricus bisporus, sterilized	2.0
Agaricus bisporus, dried	0.8
Agaricus bitorquis, fresh	18
Boletus edulis, canned	1.6
Boletus edulis, dried	0.02
Calvatia gigantea, fresh	190
Cantharellus cibarius, canned	0.15
Coprinus comatus, fresh	1.6
Gyromytra esculenta, dried	0.03
Lactarius sanguifluus, fresh	1.4
Lentinus edodes, dried	0.4
Marasmius scorodonius, dried	0.03
Pholiota squarrosa, fresh	22
Pleurotus ostreatus, fresh	17
Tricholoma nudum, fresh	0.44
Tricholoma portentsum, canned	1.8

SOURCE: After Dijkstra (1976).
NOTE: Notice the high concentrations for *Calvatia gigantea.*

While most of the flavor substances described so far in fresh and dried mushrooms have been studied only qualitatively, the metabolic pathways that lead to the biosynthesis of these compounds remain to be elucidated.

The production of flavor rich mushroom mycelium by the submerged culture process has been attempted again in the 1970s. LeDuy and coworkers (1974) examined the mycelium of a number of morel cultures cultivated in waste sulfite liquor and corn steep liquor. Intense taste and aroma characteristics could also be assured in freeze-dried powder samples of MMM (morel mushroom mycelium).

The chemical composition of the aroma substances in mycelia of *A. campestris, Coprinus comatus,* and other fungi cultivated by the submerged culture process was thoroughly analyzed by Dijkstra (1976). It was shown that 1-octen-3-ol, the major flavor component of fresh mushrooms, can also be found under submerged conditions, although incubation times were rather extended (up to twenty-eight days). There was considerable variety in the production of this compound. While the morels, *Mo. esculenta* and *Mo. hortensis* excreted most of the compound into the medium, other fungi, e.g. *A. bisporus, Co. comatus,* accumulated most of the alcohol in the mycelium. On a dry weight basis, submerged mycelium would produce more 1-octen-3-ol than the fresh mushroom in the case of *A. bisporus,* while the opposite was true with other fungi (*Co. comatus, Ln. edodes,* etc.).

Some of the compounds found in fresh mushrooms could not be detected in submerged mycelium (1-octanol, 3-octanol). It is possible that these small differences in the composition of volatile aroma compounds are related to the slight but significant difference in the flavor of submerged mycelium and fresh mushrooms. Interestingly, these two compounds were described in most of the mycelial fungi *(Aspergillus ochraceus, As. oryzae, As. parasiticus, Penicillium chrysogenum, P. citrinum, P. viridicatum)* when grown on wheat meal (Kaminski et al., 1974).

AROMA COMPOUNDS FROM OTHER MICROBES

While most of the flavor compounds of microbial origin so far discussed are produced by microorganisms related to edible foods or to food production, a variety of aroma compounds are synthesized by microbes not concerned with foods. In the present section, we shall briefly mention a number of microbial metabolites that may serve as potential sources for aroma compounds for various purposes (see also Hutchinson, 1971).

The bacterium *Pseudomonas aeruginosa,* when grown on a peptone-containing medium, produces a typical odor referred to as a grapelike sweetish aromatic, pleasant, or jasminelike odor (Mann, 1966). This was found to be due to the production of 2-amino acetophenone:

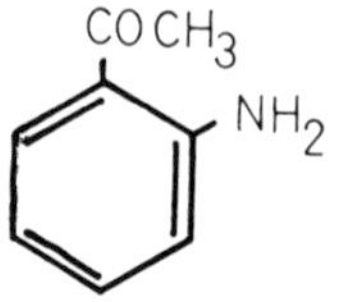

and is characteristic to this species. Related species like *Ps. fluorescens* would not produce this compound. The production of the fruity aroma by bacteria such as *Ps. fragi* has been discussed in Chapter 3.

Sprecher and coworkers (1975) describe the formation of volatile terpenes in mycelial fungi. The widely studied ascomycete *Ceratocystis coerulescens* (especially mutant 765) yielded, in a stationary culture (glucose-asparagine-malt extract), a number of terpenoids and terpenes such as 6-methyl-5-heptene-2-on, linalool, citronellol, α-terpineol, nerol, geraniol, nerolidol, farnesol. Some of these compounds appear as acetates. Similar compounds were found in the culture of *Ce. variospora* (Collins and Halim, 1970; Hubbal and Collins, 1978). Monoterpenes such as citronellol, linalool, and geraniol were also described in the yeast *Kluyveromyces lactis,* grown under aerobic, submerse conditions (Drawert and Barton, 1978).

The yeast *Sporobolomyces odorus* was found to produce the following lactones: 4-decanolide and cis-6-dodecen-4-olide, responsible for a peachlike odor (Tahara et al., 1973).

The examination of fungi for the production of new flavor compounds has opened up new possibilities in the search for microbial flavor metabolites. The fungus (basidiomycete) *Trametes odorata* produces methyl phenylacetate, which possesses a pleasant, honeylike odor, as well as methyl-p-methoxy phenylacetate, having a pleasant aniselike odor (Halim and Collins, 1971). Species of the wood-rotting fungus *Phellinus (fomes)* were found to produce a fruity aroma, which was due to methyl benzoate as well as some benzyl alcohol and methyl salicylate. In some cases (*Ph. ignarius* and *Ph. tremulae*), the sweetish linalool could also be detected (Collins and Halim, 1972). A coconutlike aroma was analyzed with cultures of the common soil fungus *Trichoderma viride* and was found to be due to 6-pentyl 2-pyrone. After Collins and Halim (1971):

6 - PENTYL - 2 - PYRONE

Volatile constituents in cultures having a "fruity-banana," "peach-pear," and "citrus" aromas were described in cultures of "*Ce. moniliformis*" (Lanza et al., 1976).

The basidiomycete *Clitocybe illudens*, when grown on agar, produces crystals directly on the mycelial mat. These crystals have been identified as the sesquiterpene torreyol (Nair and Anchel, 1973) (Fig. 34). In the basidiomycete *Mycoacia uda*, the odorous component was found to consist of p-tolualdehyde,

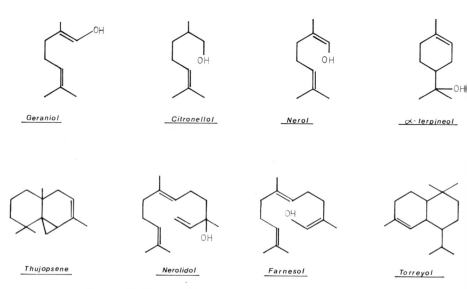

Geraniol   Citronellol   Nerol   cx-terpineol

Thujopsene   Nerolidol   Farnesol   Torreyol

Figure 34. Terpenoids and Terpenes of Fungal Origin.

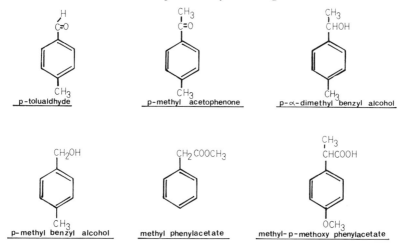

Figure 35. Odoriferous Compounds of Fungal Origin. After Halim and Collins (1975).

p-methylacetophenon, p-dimethyl benzyl alcohol and p-methylbenzyl alcohol (Halim and Collins, 1975) (Fig. 35).

The distinctive odor suggestive of a soap perfume ascribed to the activities of *Penicillium decumbens* was found to consist of a number of C8 compounds, phenethyl alcohol as well as nerolidol and thujopsene (Halim et al., 1975) (see Fig. 34).

While many of the aroma compounds have been described in fungal cultures and many more will possibly be described in the future, it seems that in most cases, terpenoids are produced by molds after prolonged incubation, usually as surface cultures. It is not excluded that these compounds are synthesized during the later phases of mycelial growth and are somehow related to the process of spore formation, similarly to the production of flavor components of *P. roqueforti* in cheeses (see Chapter 3). Quantitative data on the production of these compounds are still scarce and much greater efforts will have to be made before the production of such compounds may also be realized under submerged culture conditions.

FLAVOR ENHANCEMENT BY FOOD YEASTS

Interest in the production of food yeasts stemmed from a number of reasons. First, the possibility of employing a fast growing organism for the production of a comparatively cheap, high protein, vitamin rich preparation for a nutritive food supply (in highly populated countries or during an emergency) seemed rather promising. Second, the reduction of the high biological oxygen demand (BOD) of various industrial wastes, e.g. sulfite waste liquor of the paper industries, by growing a suitable yeast *(Candida-Torula utilis)* became an ecological necessity. Further, the possibility of employing yeasts for the production of single cell protein (SCP) on various hydrocarbons, using specially adapted yeast strains *(C. lipolytica, C. tropicalis, C. intermedia, C. guilliermondii),* raised new hopes for the use of these organisms. Although many processes have been developed with competitive technologies (Lichtfield, 1977), experimental feedings convinced experts on nutrition that even if suitable procedures became available for the acceptance of single cell protein foods, it would be advisable not to use food yeast on a large scale because of health hazards (yeasts from hydrocarbons) and the very high nucleic acid content of rapidly growing organisms which limits its use to a small fraction of man's diet. Nevertheless, research on the uses of food yeast has continued. A patent has been granted for the production of a beef flavoring agent prepared from the extract of the yeast *Pichia etchelsii* (Meat-like Flavor, 1976). Recently, the utilization of *Torula* yeasts for the improvement of flavor and palatability for special purposes, such as low calorie salad dressing and dips, has been suggested (Torutein, 1977).

Torutein, produced by a pure culture fermentation employing food grade ethanol, is said to mimic the mouth sensation of oil withoug adding calories. It is also claimed to suppress the "acid bite" of vinegar without impairing the vinegar flavor. Torutein can be further used to enhance meat foods and is particularly useful for ground meat products because it also covers soybean flavor. Its high lysine content (6.7 g/16 g N) makes it a natural supplement for corn and wheat based cereals.

MONOSODIUM GLUTAMATE AS FLAVOR ENHANCER

Many food manufacturers, especially those concerned with the production of instant soups and various ready sauces, have become aware of the necessity to strengthen the flavor of these products. Preparations of vegetable origin or with low meat ingredients require suitable meat flavor to assure acceptance by the consumer. Although the definition of the meaty taste in terms of the four basic qualities is still a matter of controversy, it is well established that the incorporation of certain amino acids into such foods considerably enhances the meat sensation by the consumer. Monosodium glutamate (MSG) is the best known example of an amino acid used as a flavor enhancer as can be seen from the steady increase in its world consumption (over 150,000 tons per year, Kinoshita and Nakayama, 1978).

The first to realize the potential use of MSG as a flavor enhancer was probably Ikeda, who succeeded at the beginning of this century to isolate MSG from the sea tangle (*Laminaria* sp.). He attributed to MSG the quality of "deliciousness" and recommended it for use as a seasoning agent (Tsuzuki and Yamashita, 1968). Interestingly, MSG is called *ajinomoto* in Japanese, the name of a company that today is a major manufacturer of this amino acid.

Glutamic acid is a common constituent of both plant and animal protein and can be isolated after suitable hydrolysis. However, from an economical point of view it is not a feasible process for a product to be employed in high concentrations in comparatively cheap foods. Monosodium glutamate was obtained for many years as by-product in the sugar industry where molasses, yielding sugar via the so-called Steffen's process, was found to be a good source of L-glutamate. However, demand greatly exceeded the supply so that new sources had to be developed. A major breakthrough was the discovery by a Japanese group of certain bacteria capable of producing large amounts of this amino acid (Kinoshita, 1959).

The microbial production of glutamic acid has the obvious advantage of synthesizing the L-isomer since the D-acid occurring in any organic synthesis is devoid of the flavor-enhancing properties. The best known of the glutamic acid bacteria is *Corynebacterium glutamicum* also known as *Micrococcus glutamicus*. Under suitable fermentation conditions, a nutrient medium containing a number of inorganic salts and a suitable carbon source, this organism is capable of producing large amounts of L-glutamic acid (30 to 50 g/liter and above), most of which is excreted into the medium. What makes this organism such an efficient converter of the carbon skeleton of a carbohydrate into glutamic acid? Two major reasons are given for this peculiar behavior. First, all the strains found suitable for glutamate production have a defective TCA cycle with very poor α-ketoglutarate dehydrogenase activity. The conversion of α-ketoglutarate to succinate is therefore sluggish, leading to an accumulation of this keto acid which is converted to glutamate by reductive animation (Fig. 36).

The absence of a regulatory mechanism that would inhibit the accumulation of such an end product can be explained by another interesting feature of these organisms. Glutamic acid producing bacteria are biotin-requiring organisms. In the absence of suitable amounts of this vitamin, the cytoplasmic membrane of the bacterium loses much of its barrier properties, permitting the leakage of the amino acid produced within the cell into the environment. This has been elegantly demonstrated by Takinami et al. (1966) (Fig. 37). The manufacture of glutamic

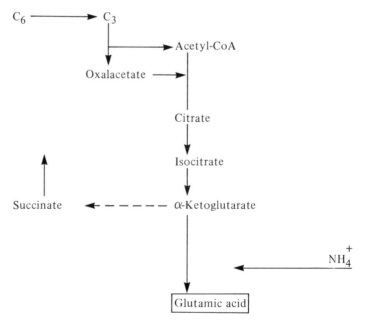

Figure 36. Major Steps in the Synthesis of Glutamic Acid.

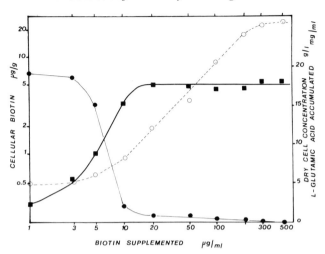

Figure 37. The Effect of Biotin on the Accumulation of Glutamic Acid by *Brevibacterium lactofermentum*. After Takinami et al. (1966). Open circle: cellular biotin, closed circle: L-glutamic acid excreted, rectangle: dry cell weight.

acid by a fermentation process in which the level of biotin is suitably controlled makes this flavor-enhancing amino acid available in practically unlimited quantities. However, it has been recently shown that temperature sensitive mutants of *Brevibacterium lactofermenti* (strain 2256, Ts-88) may produce large amounts of glutamic acid even in the presence of biotin rich media when fermented at the nonpermissive temperature (Momose and Takagi, 1978).

Further information on the biochemistry of glutamate synthesis may be found in the review by Demain and Birnbaum (1968). Additional data and recovery technology are available in the review by Powell (1968).

Pure monosodium glutamate has a distinct taste — a pleasant, mild flavor with a persistent sweet and salty taste and some tactile sensation. It has a threshold value of about 300 ppm in water. Its flavor-enhancing effect may be noticed at much lower concentrations, 150 ppm for chicken bouillon and even 20 ppm for carbonated beverages. It is also effective in depressing a variety of off-flavors caused by certain sulfur-containing compounds (Fenaroli, 1975).

Other amino acids also seem to be potential flavor enhancers.

Japanese workers have described two amino acids isolated from the mushrooms *Tricholoma muscarium* and *Amanita strobiliformis*, named tricholomic acid and ibotenic acid respectively:

$$H_2C \quad CH \; CH \; COOH \qquad HC \quad C \; CH \; COOH$$

TRICHOLOMIC ACID   IBOTENIC ACID

These amino acids have fairly low threshold values (10 to 30 ppm) and seem to considerably potentiate the flavor-enhancing effect of MSG (Kuninaka, 1966). So far, there have been no reports on the commercial production or utilization of these compounds.

NUCLEOTIDES AS FLAVOR POTENTIATORS

The search for flavor potentiators has by no means been limited to amino acids. As early as 1913, Kodama pointed out that certain nucleotides have strong flavor-enhancing properties. Thus, nucleotides have become of interest not only to the molecular biologist but also to the food technologist. Research on the chemical structure of nucleotides and their flavor potential has revealed a number of interesting features (Fig. 38). First, it must contain a purine base, hydroxylated in the 6 position. The nature of the sugar moiety is less important, although ribose is preferable. Second, the phosphate ester in position 5' yields the highest flavor activity. Di- and triphosphates are less powerful than the monophosphate. With regard to position 2 of the purine base there are three possibilities (see Fig. 38). GMP is the most powerful flavor potentiator, IMP the second best. When comparing the flavor thresholds of GMP to monosodium glutamate, the former is about ten times more powerful. In addition, there is a strong synergism between the two compounds (Fenaroli, 1975). Thus, in a 0.1 percent solution of MSG, the threshold value of GMP, which is readily distinguishable from pure MSG, drops to about 0.3 ppm, thus yielding an ideal formula for practical purposes ("potentiating the potentiator"). Indeed, GMP and IMP are admixtured with MSG at the level of about 1 percent to 10 percent in many food preparations. The nucleotides have special properties of their own. They create a sense of greater apparent viscosity in liquid foods. Soups assume more body or, as called in the trade, "greater mouth feel."

All three nucleotides concerned with flavor enhancement occur in nature. However, only GMP takes part in the structure of ribonucleic acid, while IMP and XMP appear in the metabolic pool as intermediates for the synthesis of various nucleotides. Many foodstuffs contain considerable quantities of flavor nucleotides, and it is very likely that the specific taste characteristic of certain foods is at least due in part to their nucleotide content. Meat and fish contain large amounts of nucleotides, mostly IMP which is found at concentrations of several hundred mg/g muscle tissue (Spinelli, 1967). The nucleotide content of foods and much of their flavor change greatly according to their processing. An interesting case is the development of meat flavor. Meat prepared from the freshly slaughtered animal has only poor taste. It is common practice to keep meat for a short time at refrigeration temperatures (usually 12 hours) before use. During this time, the flavor of meat reaches its maximum (Dannert and Pearson, 1967). This is due to the change in nucleotide composition during ripening. Freshly slaughtered muscles are very rich in ATP, which contributes little if at all to meat flavor. After slaughter, autolytic enzymes degrade ATP to AMP. This is followed by a deamination at position 6 and formation of IMP, which imparts the strong meaty taste. After additional storage, IMP is further degraded by muscle enzymes to hypoxanthine,

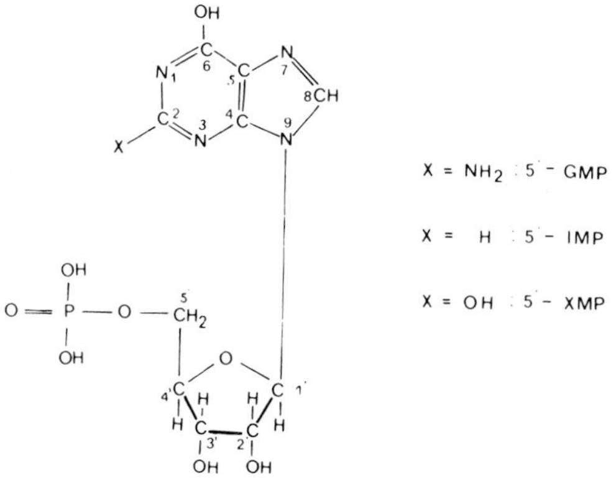

$X = NH_2 : 5' - GMP$

$X = H : 5' - IMP$

$X = OH : 5' - XMP$

Figure 38. Structure of Flavor-enhancing Mononucleotides.

which leads to a loss of the desirable flavor and the formation of a somewhat bitter taste:

$$ATP \rightarrow AMP \rightarrow IMP \rightarrow \text{Hypoxanthine}$$

PRODUCTION OF FLAVOR NUCLEOTIDES

5' mononucleotides may be prepared from muscle tissue rich in IMP. However, this is not common practice because of technical and economic reasons. GMP is found at relatively high concentrations in the mushroom *Cortinellus shiitake* (Shimazono, 1965). Yeast RNA is an excellent source for nucleotides. Of the four nucleotides set free after suitable enzymatic hydrolysis, only GMP is valuable. GMP may be recovered by a number of steps involving activated charcoal and ion exchange resins. Of the remaining three nucleotides, only AMP may be further transformed into IMP by suitable deamination — enzymatic or chemically by nitrite (Ogata et al., 1976). Another approach is the production of nucleotides by a number of microorganisms under a variety of physiological stresses. Since microbial cells are very rich in RNA (about 20% of the dry weight, mainly ribosomal RNA), certain treatments such as chilling, heating, irradiation, and antibiotics, will induce endocellular nucleases to degrade certain RNA fractions that will leak into the environment (Demain, 1969).

However, the most useful procedure for the production of flavor nucleotides would be the direct fermentation of a pure culture capable of producing and accumulating the desired nucleotides. This has become a reality after the basic steps for the biosynthesis of these nucleotides have been studied and methods for the regulation of the biosynthetic machinery to suit our needs have become available (Demain, 1968, 1978).

Biosynthesis of purine ribonucleotides begins with 5-phospho-α-D-ribosyl pyrophosphate (PRPP), which after ten steps is transformed into inosine-5'-monophosphate which constitutes the branching point for the biosynthesis of AMP and GMP (Fig. 39). While AMP and GMP control and regulate the activity of their respective enzymes, adenylosuccinate synthetase and IMP dehydrogenase, both end products, also affect (though with different intensity) the activity of earlier enzymes in the pathway — mainly PRPP amidotransferase (Ishii and Shiio, 1973).

In order to isolate a new microbe to be used for the production of a certain chemical compound by fermentation, it is common practice to screen a large number of microorganisms from culture collections and from natural sources. A similar approach was adopted by a number of American and Japanese workers, which resulted in the isolation of certain bacterial strains capable of accumulating mononucleotides or mononucleosides in large amounts which are then excreted into the medium. Among these, certain strains of *B. subtilis, Brevibacterium ammoniagenes,* the glutamic acid bacterium *C. glutamicum,* and also some strains of the streptomyces group, known for their ability to produce antibiotics, have been found useful (Nakayama et al., 1965; Demain, 1966; Misawa et al., 1969; Schwartz and Margalith, 1972).

In order to induce these organisms to produce high amounts of flavor nucleotides, it is necessary to overcome some of the control mechanisms that regulate the enzymic activity of these pathways. This can be achieved by a number of ways. First, suitable mutants can be prepared that do not produce enzymes that draw on specific intermediates of the flavor branch for the production of nucleotides not concerned with flavor. Second, the removal of the end product GMP from the system keeps the machinery at its maximum rate; and third, changing the sensitivity of the key enzyme IMP dehydrogenase and making it more tolerant to higher concentrations of the end product GMP.

All these approaches have been attempted in different laboratories, and much information has accumulated in this field of research. Mutagenic work, i.e. the preparation of strains defective in their metabolic pathways by irradiation or mutagenic chemicals, has resulted in a number of strains blocked between IMP and adenosylosuccinate (see Fig. 39). Since AMP is a strong feedback inhibitor of the key enzyme PRPP amidotransferase, which is much less sensitive to GMP, an auxotroph that is deficient of the adenylosuccinate synthetase will accumulate IMP or the nucleoside inosine (many organisms cannot excrete the nucleotide but can excrete the nucleoside), provided

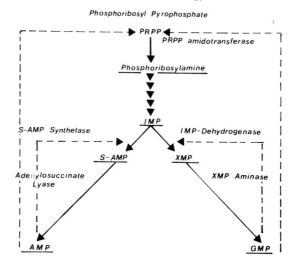

Figure 39. Biosynthesis of Flavor Nucleotides. After Demain, 1968.

very small amounts of adenosine (rather than AMP) are supplied in the medium, sufficient to supply the minimal requirements for the growth of such a mutant. Thus, production of inosine or IMP in gram quantities per liter medium can be realized. Such organisms are still subject to a certain degree of GMP depression of synthesis of the enzymes of the common pathway. To minimize the severity of this regulation, the adenine auxotrophs have been further mutated to delete the ability to synthesize IMP dehydrogenase. Such a double, adenine-xanthine auxotroph of *B. subtilis* indeed accumulates up to 15 g/liter of inosine under conditions of limited adenine and xanthine supply (Momose and Shiio, 1969). A similar approach has been employed for the efficient production of xanthine or XMP. A mutant deficient of XMP-aminase will accumulate this nucleoside in gram quantities (Ishii and Shiio, 1973).

The development of a direct fermentation to yield GMP has been more difficult than with the intermediary compounds IMP or XMP or their respective nucleosides. The activity of IMP dehydrogenase, which mediates the transformation of IMP into XMP, is very sensitive to GMP regulation. Some success has been achieved by using an adenine auxotroph in which the IMP

dehydrogenase has been desensitized by a suitable back mutation. Such a strain of *C. glutamicum* would yield up to 1 g/liter of
GMP (Demain et al., 1966). Similar yields of GMP could be
obtained by a *Streptomyces* sp. strain that was grown in the presence of a suitable anion exchange resin or by careful addition to
the medium of dioxan (Schwartz and Margalith, 1973). The
latter compound acts by quantitatively forming a complex with
GMP and water, thus relieving the feedback inhibition of the
sensitive IMP dehydrogenase. Further improvements could be
achieved by the selection of strains made resistant to certain
antimetabolites, like 8-azaguanine, which resulted in organisms
that apparently contained an IMP dehydrogenase much less
sensitive to the end product inhibition by GMP. Increased production of guanosine by a modified auxotroph of *B. subtilis*
(AJ-11) has been claimed more recently (Shibai et al., 1978). The
fermentation of the nucleoside (up to 20 g/liter) may be followed
by a chemical phosphorylation step for the production of
5'-GMP.

It seems that the production of the major flavor-enhancing
mononucleotides by suitable fermentation processes in gram per
liter quantities can now be realized on an industrial scale. Thus, a
new array of fermentation products has been added to the
fermentation line concerned with the production of the major
flavor-enhancing monosodium glutamate.

BIBLIOGRAPHY

Collins, R. P. and Halim, A. F.: Production of monoterpenes by the filamentous
    fungus *Ceratocystis variospora*. *Lloydia, 33:*481, 1970.
Dannert, A. D. and Pearson, A. M.: Concentration of inosine-5'-mono-
    phosphate in meat. *J Food Sci, 32:*49, 1967.
Demain, A. L.: Fermentation process for producing guanosine 5'-phosphates.
    U.S. Pat. 3,269,916, 1966.
———: Production of purine nucleotides. *Prog Ind Ferment, 8:*35, 1968.
———: Production of nucleotides by microorganisms. In Rose, A. M. (Ed.):
    Economic Microbiology, vol. 2. London, New York, San Francisco, Acad Pr,
    1978.
Demain, A. L. and Birnbaum, J.: Alteration of permeability for the release of
    metabolites from the microbial cell. *Curr Topics Microbiol Immunol, 46:*1,
    1968.

Demain, A. L., Jackson, M., Vitali, R. A., Hendlin, D., and Jacob, T. A.: Production of guanosine-5'-monophosphate and inosine-5'-monophosphate by fermentation. *Appl Microbiol, 14:*821, 1966.

Dijkstra, F. Y.: *Submerged Cultures of Mushroom Mycelium as Sources of Protein and Flavour Compounds.* Doctoral thesis, Delft, 1976.

Dijkstra, F. Y. and Wiken, T. O.: Studies on mushroom flavors. 2. Flavor compounds in *Coprinus comatus. Z Lebensm Unters Forsch, 160:*263, 1976.

Drawert, F. and Barton, H.: Biosynthesis of flavor compounds by microorganisms. 3. Production of monoterpenes by the yeast *Kluyveromyces lactis. J Agric Food Chem, 26:*765, 1978.

Eddy, B. P.: Production of mushroom mycelium by submerged cultivation. *J Sci Food Agric, 9:*644, 1958.

Fenaroli, G.: *Handbook of Flavor Ingredients,* 2nd ed. Cleveland, Ohio, CRC Pr, 1975.

Gilbert, F. A.: The submerged culture of *Morchella. Mycologia, 52:*201, 1960.

Gray, W. D.: The use of fungi as food and in food processing. *CR Food Technol, 3:*121, 1972.

Halim, A. F. and Collins, R. P.: An analysis of the odorous constituents produced by various species of *Trametes odorata. Lloydia, 34:*451, 1971.

———: Characterization of the major aroma constituents of the fungus *Mycoacia uda. Lloydia, 38:*87, 1975.

Halim, A. F., Narciso, J. A., and Collins, R. P.: Odorous constituents of *Penicillium decumbens. Mycologia, 67:*1158, 1975.

Hubbal, J. A. and Collins, R. P.: A study of factors affecting the synthesis of terpenes by *Ceratocystis variospora. Mycologia, 70:*117, 1978.

Hutchinson, S. A.: Biological activity of volatile fungal metabolites. *Trans Brit Mycol Soc, 57:*185, 1971.

Ishii, K. and Shiio, I.: Regulation of purine nucleotide biosynthesis in *Bacillus subtilis. Agric Biol Chem, 37:*287, 1973.

Kaminski, E., Stawicki, S., and Wasowicz, E.: Volatile flavor compounds produced by molds of Aspergillus, Penicillium, and Fungi imperfecti. *Appl Microbiol, 27:*1001, 1974.

Kinoshita, S.: The production of amino acids by fermentation processes. *Adv Appl Microbiol, 1:*201, 1959.

Kinoshita, S. and Nakayama, K.: Amino acids. In Rose, A. M. (Ed.): *Economic Microbiology,* vol. 2. London, New York, San Francisco, Acad Pr, 1978.

Kodama, S.: On the separation of inosinic acid. *J Chem Soc, 34:*1751, 1913.

Kuninaka, A.: Recent studies of 5' nucleotides as new flavor enhancers. In Gould, R. F. (Ed.): *Flavor Chemistry. Adv Chem Series, 56:*261, 1966.

Lanza, E., Ko, K. H., and Palmer, J. K.: Aroma production by cultures of *Ceratocystis moniliformis. J Agr Food Chem, 24:*1247, 1976.

LeDuy, A., Kosaric, N., and Zajic, J. E.: Morel mushroom mycelium growth in waste sulfite liquors as source of protein in flavoring. *Can Inst Food Sci Technol J, 7:*44, 1974.

Lichtfield, J. H.: Morel mushroom mycelium as a food-flavoring material. *Biotechnol Bioeng, 9:*289, 1967.

———: Use of hydrocarbon fraction for the production of SCP. *Biochem Bioeng Symp, 7:*77, 1977.

Mann, S.: Über dem Geruchsstoff von *Pseudomonas aeruginosa. Arch Mikrobiol, 54:*184, 1966.

Meat-like flavor. Jap. Pat. 5,132,704, 1976.

Misawa, M., Nara, T., and Kinoshita, S.: Production of nucleic acid related substances by fermentation processes. *Agric Biol Chem, 33:*514, 1969.

Momose, H. and Shiio, I.: Genetic and biochemical studies on 5'-nucleotide fermentation. *J Gen Appl Microbiol, 15:*399, 1969.

Momose, H. and Takagi, T.: Glutamic acid production in biotin-rich media by temperature-sensitive mutants of *Brevibacterium lactofermentum,* a novel fermentation process. *Agric Biol Chem, 42:*1911, 1978.

Morita, K. and Kobayashi, S.: Isolation and synthesis of lenthionine, an odorous substance of Shiitake, an edible mushroom. *Tetrahedron Lett,* 1966, p. 573.

Nair, M. S. and Anchel, M.: Metabolic products of *Clitocybe illudens. Lloydia, 36:*106, 1973.

Nakayama, K., Nara, T., Tanaka, H., Sato, Z., Misawa, M., and Kinoshita, S.: Production of nucleic acid related substances by fermentation processes. *Agric Biol Chem, 29:*234, 1965.

Ney, K. H. and Freytag, W. G.: Champignon-Aromen, Organoleptik von Strukturanalogen des 1-octene-3-ol. *Gordian, 78:*144, 1978.

Ogata, K., Kinoshita, S., Tsunoda, T., and Aida, K.: *Microbial Production of Nucleic Acid Related Compounds.* New York, Wiley, 1976.

Peppler, H. J.: *Microbial Technology.* New York, Reinhold, 1967.

Powell, R.: Monosodium glutamate and glutamic acid. *Chem Proc Review, 25:*59, 1968.

Pyysalo, H.: Identification of volatile compounds in seven edible fresh mushrooms. *Acta Chem Scand B, 30:*235, 1976.

Pyysalo, H. and Niskanen, A.: On the occurrence of N-methyl-N-formylhydrazones in fresh and processed false morel, *Gyromitra esculenta. J Agr Food Chem, 25:*644, 1977.

Pyysalo, H. and Suhiko, M.: Odour characterization and threshold values of some volatile compounds in fresh mushrooms. *Lebensm Wissensch u Technol, 9:*371, 1976.

Robinson, R. F. and Davidson, R. S.: The large-scale growth of higher fungi. *Adv Appl Microbiol, 1:*261, 1959.

Schwartz, J. and Margalith, P.: Production of flavor-enhancing materials by streptomycetes. *J Appl Bacteriol, 35:*271, 1972.

———: Binding of end product in the fermentation of nucleotides. *Biotechnol Bioeng, 15:*85, 1973.

Shibai, H., Enei, H., and Hirose, Y.: Purine nucleoside fermentation. *Proc*

*Biochem, 13:*6, 1978.

Shimazono, H.: A new flavor enhancer, the 5'-ribonucleotides. *Food Manufacture, 40:*11, 53, 1965.

Singer, R.: *Mushrooms and Truffles.* London, Leonard Hill, Ltd, 1961.

Spinelli, M.: Degradation of nucleotides in ice stored halibut. *J Food Sci, 32:*38, 1967.

Sprecher, E., Kubeczka, K. H., and Ratschko, M.: Flüchtige Terpene in Pilzen. *Arch Pharm, 308:*843, 1975.

Sugihara, T. F. and Humfeld, M.: Submerged culture of mycelium of various species of mushroom. *Appl Microbiol, 2:*170, 1954.

Sugimori, T., Oyama, Y., and Omichi, T.: Studies on basidiomycetes. 1. Production of mycelium and fruiting body from non carbohydrate organic substances. *J Ferm Technol, 49:*435, 1971.

Szuecs, J.: Method of enhancing mushroom mycelium flavor. U.S. Pat. 2,693,664, 1954.

Tahara, S. Fujinara, K., and Mizutani, J.: Neutral constituents of volatiles in cultured broth of *Sporobolomyces odorus. Agr Biol Chem, 37:*2855, 1973.

Takinami, K., Yamada, Y., and Okada, H.: Biochemical effects of fatty acids and its derivatives on L-glutamic acid fermentation. IV. Biotin. *Agric Biol Chem, 30:*674, 1966.

Thomas, A. F.: An analysis of the flavor of the dried mushroom, *Boletus edulis. J Agric Food Chem, 21:*955, 1973.

Torutein. Technical information, Amoco Food Co, 1977.

Tsuzuki, Y. and Yamashita, A.: History of the chemistry of taste in Japan. *Jap Studies History Sci, 7:*1, 1968.

Underkoffler, L. and Hickey, R. J.: *Industrial Fermentations.* New York, Chem Pub, 1954.

Varoquaux, P., Dubois, P., Avisse, C., and Rigaud, J.: Enzyme formation of the flavor of the cultivated mushroom (*Agaricus bisporus,* Linnaeus). *Riv Ital Ess Prof, 59:*182, 1977.

Worgan, J. T.: Culture of the higher fungi. *Prog Ind Microbiol, 8:*73, 1968.

AUTHOR INDEX

279

SUBJECT INDEX

A

291

ORGANISM INDEX

A

ACTINOMYCETALES, 6, 14, 43, 134, 135
Acetobacter, 5, 179, 183
Achromobacter, 85, 120, 247
  lipolyticum, 69
Aerobacter
  aerogenes, 49
Agaricus, 7
  bisporus, 259-262
  bitorquis, 261
  campestris, 257, 258, 262
Alcaligenes, 85, 120
Amanita
  strobiliformis, 270
ASCOMYCETES, 1, 7, 173
Aspergillus, 7, 93, 132
  amstelodami, 249
  fumigatus, 249
  niger, 94, 256
  ochraceus, 262
  oryzae, 149, 151, 170, 241, 262
  parasiticus, 262
  soyae, 149

B

Bacillus, 6
  amaracrylus, see B. polymyxa
  coagulans, 168
  fitzianus, 30
  laterosporus, 168
  natto, 87, 151
  polymyxa, 250
  subtilis, 87, 120, 169, 273, 274, 275
Bacterium (see also Brevibacterium)
  acetylcholini, see Lactobacillus plantarum
  esteroaromaticum, 30
BASIDIOMYCETES, 9, 263, 264
Betacocccus, see Leuconostoc
Bifidobacterium
  bifidus, 43
Boletus, 7
  edulis, 259-261
Botrytis
  cinerea, 203-206, 212

Brettanomyces, 182, 241
  bruxellensis, 175, 240
  lambicus, 240
Brevibacterium
  ammoniagenes, 273
  erythrogenes, 91, 97
  lactofermenti, 269
  lactofermentum, 269
  linens, 91, 94, 97

C

Candida, 169, 174, 266
  albicans, 185
  globiformis, 86
  guilleromondi, 266
  intermedia, 266
  krusei, 148
  lipolytica, 126, 127, 266
  parapsilosis, 148, 215
  pelliculosa, 148
  pseudotropicalis, 60, 86, 97
  pulcherrima, 175, 182, 215
  tropicalis, 266
  utilis, 148
Calvatia
  gigantea, 261
Cantharellus
  cibarius, 260, 261
Ceratocystis
  coerulescens, 263
  moniliformis, 264
  varispora, 263
Chlamydomonas, 11, 12
Chondrus, 13
Cladophora, 12
Cladosporium, 132
  butyri, 86
Clitocybe
  illudens, 264
Clostridium, 14
  sporogenes, 100
  tyrobutyricum, 100
  welchii, 250
Coprinus

305

Z